Physics and Ecology in Fluids

Physics and Ecology in Fluids
Modeling and Numerical Experiments

Marek Stastna
Department of Applied Mathematics
University of Waterloo
Waterloo, ON, Canada

Derek Steinmoeller
Aquanty Inc.
Waterloo, ON, Canada

Elsevier
Radarweg 29, PO Box 211, 1000 AE Amsterdam, Netherlands
The Boulevard, Langford Lane, Kidlington, Oxford OX5 1GB, United Kingdom
50 Hampshire Street, 5th Floor, Cambridge, MA 02139, United States

Copyright © 2023 Elsevier Inc. All rights reserved.

MATLAB® is a trademark of The MathWorks, Inc. and is used with permission.
The MathWorks does not warrant the accuracy of the text or exercises in this book.
This book's use or discussion of MATLAB® software or related products does not constitute endorsement or sponsorship by The MathWorks of a particular pedagogical approach or particular use of the MATLAB® software.

No part of this publication may be reproduced or transmitted in any form or by any means, electronic or mechanical, including photocopying, recording, or any information storage and retrieval system, without permission in writing from the publisher. Details on how to seek permission, further information about the Publisher's permissions policies and our arrangements with organizations such as the Copyright Clearance Center and the Copyright Licensing Agency, can be found at our website: www.elsevier.com/permissions.

This book and the individual contributions contained in it are protected under copyright by the Publisher (other than as may be noted herein).

Notices

Knowledge and best practice in this field are constantly changing. As new research and experience broaden our understanding, changes in research methods, professional practices, or medical treatment may become necessary.

Practitioners and researchers must always rely on their own experience and knowledge in evaluating and using any information, methods, compounds, or experiments described herein. In using such information or methods they should be mindful of their own safety and the safety of others, including parties for whom they have a professional responsibility.

To the fullest extent of the law, neither the Publisher nor the authors, contributors, or editors, assume any liability for any injury and/or damage to persons or property as a matter of products liability, negligence or otherwise, or from any use or operation of any methods, products, instructions, or ideas contained in the material herein.

ISBN: 978-0-323-91244-0

For information on all Elsevier publications
visit our website at https://www.elsevier.com/books-and-journals

Publisher: Candice G. Janco
Acquisitions Editor: Maria Elekidou
Editorial Project Manager: Sara Valentino
Production Project Manager: Kumar Anbazhagan
Cover Designer: Greg Harris

Typeset by VTeX

Dedicated to Martinique, Misha, and Myra (MS) and to Aria, Emilia, and Jen (DS)

Contents

List of figures xi
List of tables xxi
List of source code files xxiii
Preface xxv
Acknowledgments xxvii

CHAPTER 1 Introductory remarks and reader's guide **1**
    1.1 Introduction 1
    1.2 Reader's guide 5
          References 7

CHAPTER 2 The basics of modeling: Mechanics and population models **9**
    2.1 A (not so) basic model 9
    2.2 Exponential growth 13
    2.3 Logistic growth 14
    2.4 The Lotka-Volterra model 15
    2.5 Extending the Lotka-Volterra model 18
    2.6 Models of plankton populations 25
    2.7 Mini-projects 27
    2.8 Computing spectra 28
          References 30

CHAPTER 3 Modeling active tracers **31**
    3.1 Numerical experiments 31
    3.2 Active tracers: the simplest models 32
    3.3 Nonlinear effects 34
    3.4 Interacting tracers: the Lotka-Volterra model 36
    3.5 Advection-diffusion for plankton models 39
    3.6 Tracers in two dimensions 42
    3.7 Advection diffusion population model in 2D 46
    3.8 Mini-projects 49
          References 50

CHAPTER 4 Fluid mechanics **51**
    4.1 The basics of fluid mechanics: accounting for flow 51
    4.2 Forces in a fluid: generalizing Newton's second law 53
    4.3 The shallow water equations 57
    4.4 Vorticity and stream function 62
    4.5 Mini-projects 64
          References 64

vii

viii Contents

CHAPTER 5 **Rotating shallow water dynamics: An overview** **65**
  5.1 An overview of material 65
  5.2 The effect of rotation: geostrophy 66
  5.3 Rotating gravity waves in a channel: derivation 68
  5.4 Rotating gravity waves in a channel: discussion 76
  5.5 The circular lake: derivation 79
  5.6 The circular lake: discussion 81
  5.7 The tilted free surface problem 82
  5.8 An instability problem: simulation 85
  5.9 Mini-projects 89
    References 89

CHAPTER 6 **Rotating shallow water dynamics: Dispersion and nonlinearity** **91**
  6.1 Realism in wave models 91
  6.2 The dispersive shallow water model 92
  6.3 Adjustment problems on the field scale 96
  6.4 Seiche evolution 102
  6.5 Coding ideas 110
  6.6 Mini-projects 114
    References 114

CHAPTER 7 **Understanding complex dynamics in two and three dimensions** **115**
  7.1 Turbulence closures: the idea 115
  7.2 Turbulence closures: Navier Stokes 118
  7.3 Large-eddy simulations (or LES) 120
  7.4 Slow-fast systems 124
  7.5 Mini-projects 129
    References 131

CHAPTER 8 **Modeling motion in the vertical** **133**
  8.1 Stratified fluid dynamics in the x-z plane 133
  8.2 The idealized internal seiche: derivation 134
  8.3 The idealized internal seiche: transport 138
  8.4 The internal seiche: arbitrary stratifications 142
  8.5 The internal seiche: finite amplitude and wavetrains 147
  8.6 Nonlinearity due to biological behavior 151
  8.7 Mini-projects 163
    References 164

CHAPTER 9 **Modeling fluid transport processes with finite volume methods** **165**
  9.1 Beyond spectral methods 165
  9.2 A short history of finite volume methods 166
  9.3 Getting acquainted with finite volume methods in 1D 166

9.4	Insight from 1D energy analysis: the continuous problem	170
9.5	Insight from 1D energy analysis: the discrete problem	171
9.6	Exercises: building intuition	173
9.7	Implementation details in 1D	173
9.8	Doing it "the right way" in 2D	174
9.9	Towards harder problems	177
9.10	Well-balanced upwinding for the case of variable bottom bathymetry	179
9.11	Returning to the dispersive shallow water system	181
9.12	Towards "high-resolution" numerical schemes	183
9.13	Wrap-up	184
9.14	Mini-projects	184
	References	185

CHAPTER 10 Modeling fluid transport processes with discontinuous Galerkin methods: Background 187

10.1	An overview of material	187
10.2	Discontinuous Galerkin finite element method for dispersive shallow water equations	188
10.3	Evaluating the inner products: modes and nodes	192
10.4	Polynomial interpolation nodes in 2D	194
10.5	Local operators for the nodal approach	195
10.6	Surface integral contributions	196
10.7	Boundary conditions	197
10.8	Dealing with source terms: bathymetry and wave dispersion	197
10.9	Solving for the non-hydrostatic pressure: an elliptic problem	200
10.10	Hyperbolic theory in 1D	202
10.11	Linear Riemann problem	203
10.12	Time-stepping method	205
10.13	Mini-projects	206
	References	207

CHAPTER 11 Modeling fluid transport processes with discontinuous Galerkin methods: Implementation . 209

11.1	Towards practical implementations of DG-FEM	209
11.2	Spurious eddies in inviscid DG-FEM solutions	210
11.3	Curvilinear elements	214
11.4	Constructing coordinates systems for curvilinear elements	214
11.5	Cubature and quadrature integration	217
11.6	Internal rotating seiche simulation using curvilinear elements	219
11.7	Internal rotating seiche simulation in a real-world lake	220
11.8	Wrap-up	223
	References	223

x   Contents

CHAPTER 12 Beyond standard treatments: Flow in porous media 225

    12.1   Motivation 225

    12.2   The basic theory of porous media 225

    12.3   Flow in saturated porous media: two simple examples 228

    12.4   Flow in saturated porous media: a more complex example . . . 231

    12.5   Towards more general descriptions: unsaturated flow 233

    12.6   Mini-projects 236

    12.7   Concluding remarks 236

           References 237

Glossary 239

Index 243

List of figures

Fig. 1.1 Satellite image of a portion of Lake Erie during a toxic algae bloom. The bloom is prominent on the Canadian (northern) side of the lake and manifests in complex, yet coherent patterns. Note the clear vortex off the tip of the Point Pelee peninsula.   3

Fig. 1.2 Three shorter tracks through the book. Chapters relevant for a numerical methods focus are shown in blue (in the left panel), those relevant for a modeling focus are shown in green (in the middle panel), while those relevant for a fluid mechanics focus are shown in orange (in the right panel).   6

Fig. 2.1 The definition diagram for the mass on a spring as an example of the simple harmonic oscillator or SHO.   10

Fig. 2.2 $C(t)$ (black) and $F(t)$ (blue) versus t for the Lotka-Volterra model. Only the second half of the integration time shown. $C(0) = (0.1, 0.5, 5, 10)$ for the four panels with $C(0) = 0.1$ in the top left and $C(0) = 10$ in the bottom right.   16

Fig. 2.3 Phase space plot for the Lotka-Volterra model. Only the second half of the integration time shown. $C(0) = (0.1, 0.5, 5, 10)$ for the four panels with $C(0) = 0.1$ in the top left and $C(0) = 10$ in the bottom right.   17

Fig. 2.4 Diagram of the two island Lotka-Volterra model with only rabbits being able to cross between islands. Rabbits move from the island with higher rabbit population to the one with a lower population.   18

Fig. 2.5 Phase portraits for the two island model, experiment one (change in rabbit birth rate). $k = (0, 0.0001, 0.001, 0.01)$ for the four panels with $k = 0$ in the top left. Island 1 is shown in black, island 2 in blue.   20

Fig. 2.6 Spectra (the Power Spectral Density or PSD) for the two island model, experiment one (change in rabbit birth rate). $k = (0, 0.0001, 0.001, 0.01)$ for the four panels with $k = 0$ in the top left. Island 1 is shown in black, island 2 in blue.   21

Fig. 2.7 Phase portraits for the two island model, experiment two (change in predation rate). $k = (0, 0.0001, 0.001, 0.01)$ for the four panels with $k = 0$ in the top left. Island 1 is shown in black, island 2 in blue.   22

Fig. 2.8 Phase portraits for the time dependent birth rate model ($a = 0.4$, $T = 20$). $C(0) = (0.1, 0.5, 5, 10)$ for the four panels with $C(0) = 0.1$ in the top left and $C(0) = 10$ in the bottom right.   24

Fig. 2.9 Phase portraits for the time dependent birth rate model ($a = 0.4$, $T = 20$). $C(0) = (0.1, 0.5, 5, 10)$ for the four panels with $C(0) = 0.1$ in the top left and $C(0) = 10$ in the bottom right.   25

xii   List of figures

Fig. 2.10   Evolution of the Truscott Brindley model from four different initial phytoplankton populations. The bottom left panel reproduces their Figure 4.   27

Fig. 3.1   The evolution of an advection, reaction diffusion equation. The initial state is shown in black. Later times (in dimensionless units) are shown in red, green, and blue for $t = 0.1, 0.5$, and 1.   34

Fig. 3.2   The evolution of an advection, reaction diffusion equation shown as a space time plot.   35

Fig. 3.3   The evolution of two examples of Fisher's equation. The initial state is shown in black, with a later time in blue and red (the latter for ν reduced by an order of magnitude).   37

Fig. 3.4   Space-time, or Hovmoeller plots of C (a) and F (b) with $\nu = 10^{-3}$. Note the way in which the initial population perturbation signal slopes to the right for both the broad packet envelope and the shorter length scale packet. Red-blue bands indicate the periodic, or limit cycle behavior in time.   39

Fig. 3.5   Space-time, or Hovmoeller plots of C (a) and F (b) with $\nu = 10^{-2}$. Note the way in which the initial population perturbation signal slopes to the right for both the broad packet envelope and the shorter length scale packet. Red-blue bands indicate the periodic, or limit cycle behavior in time.   40

Fig. 3.6   Space-time, or Hovmoeller plots of P (a) and Z (b) with $\nu = 10^{-3}$. Note the way in which the initial population perturbation signal slopes to the right for both the broad packet envelope for P. Propagation in the Z field is only evident after $t = 20$.   41

Fig. 3.7   Line plots of P (a) and Z (b) with $\nu = 10^{-3}$. t increases from $t = 10$ by 10 time units.   42

Fig. 3.8   Snapshots of the concentration for the advection by a shear field with diffusion equation. (a) $C(x, y, t = 0)$, (b) $C(x, y, t = 3)$ with $\nu = 10^{-3}$, (c) $C(x, y, t = 3)$ with $\nu = 10^{-2}$.   44

Fig. 3.9   Top left panel shows the streamfunction and the initial distribution of tracer (black box). Subsequent panels show the tracer field at 0.144 dimensionless time unit increments. The tracer is capped at $C = 0.1$ and the division between the two cells is indicated by a white, vertical line. $\nu = 0.001$.   46

Fig. 3.10   Top left panel shows the streamfunction and the initial distribution of tracer (black box). Subsequent panels show the tracer field at 0.144 dimensionless time unit increments. The tracer is capped at $C = 0.1$ and the division between the two cells is indicated by a white, vertical line. $\nu = 0.01$.   47

Fig. 3.11   Top left panel shows the streamfunction and the initial distribution of tracer (black box). Subsequent panels show the population field at

List of figures **xiii**

Fig. 3.12   0.144 dimensionless time unit increments. The population is capped at $C = 0.1$ and the division between the two cells is indicated by a white, vertical line. $\nu = 0.001$.   48

Fig. 3.12   Top left panel shows the streamfunction and the initial distribution of tracer (black box). Subsequent panels show the population field at 0.144 dimensionless time unit increments. The population is capped at $C = 0.1$ and the division between the two cells is indicated by a white, vertical line. $\nu = 0.01$.   49

Fig. 4.1   Cartoon showing the difference between Eulerian and Lagrangian measurements using a weather balloon and an observation tower as a warm front propagates past the tower.   53

Fig. 4.2   Summary of boundary conditions at the free surface.   58

Fig. 4.3   Real world processes that are neglected in the derivation of the shallow water equations that are often introduced on an *ad hoc* basis.   60

Fig. 4.4   Cartoon illustrating how vorticity characterizes local rotation of fluid particles. Vorticity may differ even for particles moving along the same pathline.   63

Fig. 4.5   Cartoon illustrating how the streamfunction and velocity are related on the $x - y$ plane.   63

Fig. 5.1   Sample profiles for two Poincaré waves and the Kelvin wave for surface, or barotropic waves in a channel.   78

Fig. 5.2   Sample profiles for two Poincaré waves and the Kelvin wave for internal, or baroclinic, reduced gravity waves in a channel.   78

Fig. 5.3   Numerically computed real part of the vertical mode-1 displacement, $G(r)$, for the first six Kelvin modes of a model large circular mid-latitude lake.   81

Fig. 5.4   **Top:** Snapshots of the interface displacement η. **Bottom:** Snapshots of the passive tracer B with velocity direction field super-imposed (purple arrows).   82

Fig. 5.5   A simulation with an initial free surface tilt that is one fifth of that shown in Fig. 5.4. **Top:** Snapshots of the interface displacement η. **Bottom:** Snapshots of the passive tracer B with velocity direction field super-imposed (purple arrows).   83

Fig. 5.6   Radial transects of interface displacement for the 'one-fifth amplitude' linear case simulation, with varying angle $\theta > 0$ (w.r.t. the positive x-axis) chosen to pass through the maximum value of the interface displacement along the wall boundary. Different color curves correspond to time values $t = 0$ (blue), $t = 17.6$ h (orange), $t = 34.5$ h (green), $t = 51.7$ h (red). Subsequent curves have each been shifted up along the y-axis by 0.35 relative to the previous curve. The dashed line is a reference curve depicting the theoretical decay scale of

xiv   List of figures

	an idealized Kelvin wave trapped to the wall boundary with Rossby deformation radius $L_0 = c_0/f = 6825$ m and wall boundary at $r = a$. Here, $a = 8000$ m, $f = 7.88 \times 10^{-5}$ s^{-1}, and $c_0 = 0.50$ m s^{-1} is the linear long wave speed.	84
Fig. 5.7	Conceptual picture of the effect of nonlinearity in the shallow water equations. The regions with the faster local speeds of propagation catch up to the slower speeds leading to a nearly vertical wave front, and eventually overturning.	85
Fig. 5.8	Snapshots of the u field for the breakdown of a double-jet at times (from top to bottom) $t = 0$, 1.88 h, 3.76 h, 5.64 h.	87
Fig. 5.9	Snapshots of the B (tracer) field for the breakdown of a double-jet at times (from top to bottom) $t = 0$, 1.88 h, 3.76 h, 5.64 h.	87
Fig. 5.10	Same as Fig. 5.8 but the jet velocity has been reduced by a factor of a half.	88
Fig. 5.11	Same as Fig. 5.9 but the jet velocity has been reduced by a factor of a half.	88
Fig. 6.1	Comparison of phase speeds from the traditional shallow water mode (blue), the dispersive shallow water model (green), and the full dispersion relation from potential flow theory (red).	95
Fig. 6.2	An adjustment problem for early dimensionless time ($t = 0.12$). Black – full model, blue – model without drag, green – linearized model, red – classical SW (no dispersion). Upper panel – free surface scaled by undisturbed depth, Lower – panel u scaled by c_0.	96
Fig. 6.3	An adjustment problem for late dimensionless time ($t = 0.71$). Black – full model, blue – model without drag, green – linearized model, red – classical SW (no dispersion). Upper panel – free surface scaled by undisturbed depth, Lower – panel u scaled by c_0.	97
Fig. 6.4	The Large domain adjustment problem for early dimensionless time ($t = 0.3$). Black – full model, blue – model without drag, green – linearized model, red – no rotation ($f = 0$). Upper panel – free surface scaled by undisturbed depth, Lower – panel u scaled by c_0.	99
Fig. 6.5	The Large domain adjustment problem for medium dimensionless time ($t = 0.59$). Black – full model, blue – model without drag, green – linearized model, red – no rotation ($f = 0$). Upper panel – free surface scaled by undisturbed depth, Lower – panel u scaled by c_0.	100
Fig. 6.6	The Large domain adjustment problem for late dimensionless time ($t = 0.89$). Black – full model, blue – model without drag, green – linearized model, red – no rotation ($f = 0$). Upper panel – free surface scaled by undisturbed depth, Lower – panel u scaled by c_0.	101
Fig. 6.7	Detail of the Large domain adjustment problem for late dimensionless time ($t = 0.89$), focusing on the leading wavetrain. Black – full model,	

List of figures   xv

| | blue – model without drag, green – linearized model, red – no rotation ($f = 0$). Upper panel – free surface scaled by undisturbed depth, Lower – panel u scaled by c_0. | 102 |

Fig. 6.8   Detail of the Large domain adjustment problem for late dimensionless time ($t = 0.89$), focusing on the secondary wavetrain. Black – full model, blue – model without drag, green – linearized model, red – no rotation ($f = 0$). Upper panel – free surface scaled by undisturbed depth, Lower – panel u scaled by c_0.   103

Fig. 6.9   Schematic of wind-induced changes in a stratified lake, panel (a); and the development of internal wave trains when the wind slackens, panel (b).   103

Fig. 6.10   The seiche development problem for early dimensionless time ($t = 1.2$). Black – full model, blue – model without drag, green – linearized model, red – no rotation ($f = 0$). Upper panel – free surface scaled by undisturbed depth, Lower – panel u scaled by c_0.   104

Fig. 6.11   The seiche development problem for early dimensionless time ($t = 5.9$). Black – full model, blue – model without drag, green – linearized model, red – no rotation ($f = 0$). Upper panel – free surface scaled by undisturbed depth, Lower – panel u scaled by c_0.   105

Fig. 6.12   The seiche development problem for early dimensionless time ($t = 9.5$). Black – full model, blue – model without drag, green - linearized model, red – no rotation ($f = 0$). Upper panel – free surface scaled by undisturbed depth, Lower – panel u scaled by c_0.   106

Fig. 6.13   The seiche development problem showing differences prior to wavetrain formation; at dimensionless time $t = 3.6$. Black – full model, blue – model without drag, green – linearized model, red – no rotation ($f = 0$). Upper panel – free surface scaled by undisturbed depth, Lower – panel u scaled by c_0.   107

Fig. 6.14   The detail of the wavetrain development at dimensionless time $t = 5.9$. Black – full model, blue – model without drag, green – linearized model, red – no rotation ($f = 0$). Upper panel – free surface scaled by undisturbed depth, Lower – panel u scaled by c_0.   108

Fig. 6.15   The seiche development problem for a dimensionless time ($t = 5.9$) at which wavetrains have developed. Black – full model, blue – $v_{eddy} = 0.1$, green – $v_{eddy} = 1$, red – $v_{eddy} = 10$. Upper panel – free surface scaled by undisturbed depth, Lowers – panel u and v scaled by c_0.   109

Fig. 6.16   Detail of the wavetrains at dimensionless time ($t = 5.9$). Black – full model, blue – $v_{eddy} = 0.1$, green – $v_{eddy} = 1$, red – $v_{eddy} = 10$. Upper panel – free surface scaled by undisturbed depth, Lowers – panel u and v scaled by c_0.   110

xvi List of figures

Fig. 7.1 Narrow filtering function example. Upper panel – filtering kernel function centered at zero, lower panel – raw (blue) and filtered function (red). 121

Fig. 7.2 Wider filtering function example. Upper panel – filtering kernel function centered at zero, lower panel – raw (blue) and filtered function (red). 122

Fig. 7.3 Early time evolution of balanced and unbalanced flow. (a) Free surface height, (b) kinetic energy, (c) vorticity, (d) divergence. 127

Fig. 7.4 Medium time evolution of balanced and unbalanced flow. (a) Free surface height, (b) kinetic energy, (c) vorticity, (d) divergence. 128

Fig. 7.5 Late time evolution of balanced and unbalanced flow. (a) Free surface height, (b) kinetic energy, (c) vorticity, (d) divergence. 129

Fig. 7.6 Late time evolution of vorticity and divergence. (a), (b) Full simulation, (c), (d) balanced-only or "Slow" simulation. 130

Fig. 7.7 The spatially varying components of the Smagorinsky eddy viscosity based on the balanced-only flow at a dimensionless time $t = 1.62$. 131

Fig. 8.1 Frequencies of internal seiches in a linearly stratified "aquarium" lake. $H = 20$ m, $L = 5,000$ m, $\Delta\rho = 0.01$. (a) Frequency as a function of k for the first 100 horizontal modes with the first (blue) and fifth (red) vertical mode number, (b) frequency as a function of m for the first 100 vertical modes with the first (blue) and fifth (red) horizontal mode number. 138

Fig. 8.2 Particle locations over one period of a small amplitude mode-1 internal seiche. The streamfunction is shown shaded, with a saturation of ± 1. 140

Fig. 8.3 Particle locations over one period of a large amplitude mode-1 internal seiche. The streamfunction is shown shaded, with a saturation of ± 7.5. 141

Fig. 8.4 Particle locations over one period of a small amplitude mode-2 internal seiche. The streamfunction is shown shaded, with a saturation of ± 1. 142

Fig. 8.5 Particle locations over one period of a large amplitude mode-1 wave combined with a small amplitude mode-2 internal seiche. The streamfunction is shown shaded, with a saturation of ± 7.5. 143

Fig. 8.6 Particle locations over one period of a large amplitude mode-1 wave with combined with a small amplitude mode-2 internal seiche. The streamfunction is shown shaded, with a saturation of ± 7.5. 144

Fig. 8.7 Sample temperature and $N^2(z)$ profiles for a single pycnocline and single pycnocline with an additional background stratification. 146

Fig. 8.8 Sample mode-1 and mode-2 vertical structure functions for a single pycnocline and single pycnocline with an additional background stratification. 147

Fig. 8.9 Particle locations over one period of a small amplitude mode-1 internal seiche with variable $N^2(z)$. The streamfunction is shown shaded, with a saturation of ± 1. 148

List of figures xvii

Fig. 8.10	Particle locations over one period of a small amplitude mode-2 internal seiche with variable $N^2(z)$. The streamfunction is shown shaded, with a saturation of ± 1.	149
Fig. 8.11	The shaded density field for the nonlinear stratified internal seiche. Upper panel – initial condition, middle panel – $t = 2.5$ hours, lower panel – $t = 2.5$ hours detail of region between white lines in the middle panel.	150
Fig. 8.12	The shaded density field for the nonlinear stratified internal seiche. Upper panel – $t = 3.25$ hours, lower panel white arrow indicates mode-1 nonlinear wave train, yellow arrow indicates trailing long mode-2 wave, $t = 3.25$ hours detail of region near right wall.	151
Fig. 8.13	The shaded horizontal velocity field for the nonlinear stratified internal seiche at $t = 3.25$ hours with 4 isolines of density superimposed in white.	152
Fig. 8.14	The shaded horizontal velocity field for the nonlinear stratified internal seiche at $t = 3.25$ hours with 4 isolines of density superimposed in white, in the region near the right wall.	153
Fig. 8.15	The shaded density field for the nonlinear stratified internal seiche. Upper panel – $t = 3.5$ hours, lower panel shows details of a 1 km long region near right wall.	154
Fig. 8.16	Particle paths for four sample shear triggered particles with values of critical shear changing. Colors range from the half base critical shear (thin black), base critical shear (dark gray), double base critical shear (light gray), and four times base critical shear (thick black).	156
Fig. 8.17	Particle locations for the model with birth and death, over one period of a small amplitude mode-1 internal seiche with variable $N^2(z)$. The streamfunction is shown shaded, with a saturation of ± 1.	157
Fig. 8.18	Particle locations for the model without birth and death, with the same initial conditions as the previous figure, over one period of a small amplitude mode-1 internal seiche with variable $N^2(z)$. The streamfunction is shown shaded, with a saturation of ± 1.	158
Fig. 8.19	Diagram of gyrotaxis of an ellipsoidal particle in a shear flow.	159
Fig. 8.20	Paths of gyrotactic particles in the oscillating plate flow with four different initial orientations. Orientation is specified by angle, $\pi/4$ blue, $3\pi/4$ black, $-3\pi/4$ red, $-\pi/4$ green.	160
Fig. 8.21	Orientation of gyrotactic particles in the oscillating plate flow with four different initial orientations versus time. Orientation is specified by angle, $\pi/4$ blue, $3\pi/4$ black, $-3\pi/4$ red, $-\pi/4$ green.	161
Fig. 8.22	Paths of gyrotactic particles in the oscillating plate flow with four different initial orientations. Orientation is specified by angle, $-3\pi/4$ blue, $-3\pi/4 - 0.01$ black, $-3\pi/4 - 0.02$ red, $-3\pi/4 + 0.01$ green.	162
Fig. 8.23	Inter-particle distance between the paths in the previous figure. All curves consider the difference from the $-3\pi/4$ case, with a perturbation of -0.01 blue, -0.02 black, $+0.01$ red.	163

xviii   List of figures

Fig. 9.1   Graphical depiction of 1D grid cells lying in a line segment along the real line in the finite volume method. Ghost cells, which are used to impose boundary conditions at the first and last cell edges but are not actually a part of the computational grid, are also shown.   167

Fig. 9.2   Cartoon diagram illustrating flow between neighboring finite volume cells for tracer field T and left-to-right velocity, $u > 0$. Panel (b) shows the discrete flow stage at a "fictitious time," after the solution at the edges has been reconstructed and evolved, but prior to finite volume averaging.   169

Fig. 9.3   Upwind finite volume numerical solution to the 1D tracer equation for initial condition $T(x, 0) = \cos(2\pi x / L)$.   174

Fig. 9.4   Output from the `tracer2d.py` script: A triple-Gaussian tracer field subjected to shear flow of the form $\vec{u} = (u(y), 0)$. *Top*: u velocity field component (left), initial tracer distribution $T(x, y, 0)$ (right). *Bottom*: Tracer distribution at $t = 2636$ s (left) and $t = 5273$ s (right).   178

Fig. 9.5   **Top:** The evolution of a small-amplitude Gaussian bump initialized to propagate from right-to-left as described by the finite volume (FV) method with $N = 1024$ grid cells, and the Fourier method with $N = 256$ grid points. The vertical dashed line appearing in both panels represents the wave peak as predicted by linear theory $x^* = x_0 - \sqrt{gH}t^*$, with $t^* = 160$ s. **Bottom:** Like the top panel, but the initial condition's amplitude has been increased from 0.1 m to 3.5 m, and with successively doubled FV grids of size $N = 1024$, 2048, and 4096.   182

Fig. 9.6   A repeat view of the bottom panel of Fig. 9.5, re-centered to show the details of the wavetrains more closely.   183

Fig. 10.1   Warburton's "near-optimal" polynomial interpolation nodes on an equilateral triangle for order $N = 8$.   195

Fig. 10.2   Illustration of the Riemann problem in the xt-plane. Immediately after the initialization, a third state ($\vec{q}^*$) appears in the solution that must be determined.   204

Fig. 11.1   Snapshots of the η-field in the order $N = 4$ DG-FEM simulation of a rotating seiche on a perturbed circular domain with a re-entrant peninsula at **(a)** $t = 0$ h, **(b)** $t = 6.8$ h, **(c)** $t = 14.0$ h, **(d)** $t = 20.9$ h. Note the apparent separation eddies visible near the peninsula in panels **(b)–(d)**. This figure was published in *Ocean Modelling*, **107**, D.T. Steinmoeller, M. Stastna, K.G. Lamb, "Discontinuous Galerkin methods for dispersive shallow water models in closed basins: Spurious eddies and their removal using curved boundary methods," 112–124. © 2016 Elsevier.   211

Fig. 11.2   Cartoon diagram of potential flow around a wall corner of angle $\alpha > \pi$. Contours depict streamlines (lines where $\psi = const.$) and arrows illustrate the relative strength and direction of $-\nabla p$.   213

Fig. 11.3	Illustration of straight-sided element mesh along with a smooth representation of the boundary, the cubic spline interpolant, that will be used to produce deformed elements. This figure was published in *Ocean Modelling*, **107**, D.T. Steinmoeller, M. Stastna, K.G. Lamb, "Discontinuous Galerkin methods for dispersive shallow water models in closed basins: Spurious eddies and their removal using curved boundary methods," 112–124. © 2016 Elsevier.	215
Fig. 11.4	Diagram of the reference triangle and illustration of (r, s) coordinates. This figure was published in *Ocean Modelling*, **107**, D.T. Steinmoeller, M. Stastna, K.G. Lamb, "Discontinuous Galerkin methods for dispersive shallow water models in closed basins: Spurious eddies and their removal using curved boundary methods," 112–124. © 2016 Elsevier.	216
Fig. 11.5	**Left**: A pair of elements before being deformed. **Right**: The same elements after being deformed to match the cubic-spline representation of the boundary with interior nodes re-distributed via Gordon-Hall blending. This figure was published in *Ocean Modelling*, **107**, D.T. Steinmoeller, M. Stastna, K.G. Lamb, "Discontinuous Galerkin methods for dispersive shallow water models in closed basins: Spurious eddies and their removal using curved boundary methods," 112–124. © 2016 Elsevier.	217
Fig. 11.6	Panels **(a)**–**(d)**: Like Fig. 11.1 but with curvilinear elements along the boundary. The other panels correspond to the later times **(e)** $t = 28.1$ h, **(f)** $t = 34.9$ h, **(g)** $t = 42.1$ h, **(h)** $t = 49.0$ h. This figure was published in *Ocean Modelling*, **107**, D.T. Steinmoeller, M. Stastna, K.G. Lamb, "Discontinuous Galerkin methods for dispersive shallow water models in closed basins: Spurious eddies and their removal using curved boundary methods," 112–124. © 2016 Elsevier.	219
Fig. 11.7	Panel **(a)**: Depth (in m) of Pinehurst Lake, AB from raw 50 m bathymetry data, and panel **(b)**: corresponding $H = 0$ contour (black) with smoothed coastline super-imposed (red). The lower panels show a zoomed-in section of the **(c)** straight-sided and **(d)** curved ($N = 6$) finite element mesh with $K = 1807$ elements near $(x, y) = (7$ km$, 5$ km$)$ with cubic spline interpolant super-imposed (red). This figure was published in *Ocean Modelling*, **107**, D.T. Steinmoeller, M. Stastna, K.G. Lamb, "Discontinuous Galerkin methods for dispersive shallow water models in closed basins: Spurious eddies and their removal using curved boundary methods," 112–124. © 2016 Elsevier.	221
Fig. 11.8	Evolution of an interfacial tilt in Pinehurst Lake, AB using the $N = 6$ DG-FEM with curvilinear boundary elements at times **(a)** $t = 0$ h, **(b)** $t = 19.4$ h, **(c)** $t = 39.3$ h, **(d)** $t = 62.7$ h. This figure was published in *Ocean Modelling*, **107**, D.T. Steinmoeller, M. Stastna, K.G. Lamb, "Discontinuous Galerkin methods for dispersive shallow water models in closed basins: Spurious eddies and their removal using curved boundary methods," 112–124. © 2016 Elsevier.	222

xx List of figures

Fig. 11.9   Like Fig. 11.8, except the kinetic energy density, $\frac{1}{2}h(u^2 + v^2)$ is plotted. This figure was published in *Ocean Modelling*, **107**, D.T. Steinmoeller, M. Stastna, K.G. Lamb, "Discontinuous Galerkin methods for dispersive shallow water models in closed basins: Spurious eddies and their removal using curved boundary methods," 112–124. © 2016 Elsevier.   222

Fig. 12.1   Physical picture vs. mathematical picture of porous media description.   226

Fig. 12.2   **(a)** Porous lake bottom conceptual diagram and **(b)** laboratory demonstration photo of buoyancy driven flow over a porous bottom driving flow through the porous region.   228

Fig. 12.3   Porous lake bottom with a region magnified to show a small, semi-circular bump on the porous bottom. The flow is assumed to be uni-directional and uniform far from the bump. In the text we explain how this can facilitate an exact solution.   232

List of tables

Table 6.1	Parameters for the Adjustment case.	96
Table 6.2	Parameters for the Field Scale Adjustment cases.	97
Table 6.3	Parameters for the Field Scale Adjustment cases.	102
Table 7.1	Parameters for the Slow-Fast simulation.	125

List of source code files

File name	MATLAB	Python	Chapter
lotka_volterra_plain.m	✓		2
lotka_volterra_two.m	✓		2
lotka_volterra_twob.m	✓		2
lotka_volterra_time_dep.m	✓		2
truscott_brindley_plain.m	✓		2
adv_diff_react_1d.m	✓		3, 9
adv_diff_react_1d.py		✓	
adv_diff_fisher.m	✓		3
adv_diff_lotka_volterra.m	✓		3
adv_diff_tandb.m	✓		3
adv_diff_shear.m	✓		3
adv_diff_cells_book.m	✓		3
swnh1d_shortscale.m	✓		6
swnh1d_large_adjust.m	✓		6
swnh1d_seiche.m	✓		6
swnh1d_weddyvisc.m	✓		6
les_basics.m	✓		7
swnh2d_book.m	✓		7
plot2d.m	✓		7
plot_les_like.m	✓		7
constant_N_seiche_freq.m	✓		8
constant_N_seiche_bookpics.m	✓		8
constant_N_seiche.m	✓		8
constant_N_seiche_tracer.m	✓		8
linear_lake.m	✓		8
variable_N_seiche_bookpics.m	✓		8
variable_N_seiche_wbirthdeath.m	✓		8
simplest_particles.m	✓		8
gyrotactic_stokes.m	✓		8
tracer1d.py		✓	9
tracer2d.py		✓	9
sw1d_fwave.py		✓	9
sw_1d_nonhydro_fv.py		✓	9

xxiii

xxiv   List of source code files

File name	Brief description
`lotka_volterra_plain.m`	Solve the Lotka-Volterra (LV) Model
`lotka_volterra_two.m`	Solve the two-island LV model
`lotka_volterra_twob.m`	Solve the two-island LV model
`lotka_volterra_time_dep.m`	Solve the LV model with time dependent parameters
`truscott_brindley_plain.m`	Solve the Phyto-Zoo (PZ) Plankton Model
`adv_diff_react_1d.m`	Solve the advection-reaction-diffusion model in 1D
`adv_diff_react_1d.py`	Solve the advection-reaction-diffusion model in 1D
`adv_diff_fisher.m`	Solve Fisher's equation
`adv_diff_lotka_volterra.m`	Solve the avdection-diffusion LV model
`adv_diff_tandb.m`	Solve the avdection-diffusion PZ model
`adv_diff_shear.m`	Solve 2D advection-difussion with shear
`adv_diff_cells_book.m`	Solve advection diffusion with cells
`swnh1d_shortscale.m`	Solve the 1D SW with dispersion model (short scale)
`swnh1d_large_adjust.m`	Solve the 1D SW with dispersion model (large scale)
`swnh1d_seiche.m`	Solve the 1D SW with dispersion model (seiche)
`swnh1d_weddyvisc.m`	Solve the 1D SW with dispersion model (eddy viscosity)
`les_basics.m`	LES filtering basics
`swnh2d_book.m`	Solve the 2D SW with dispersion model
`plot2d.m`	2D plots for `swnh2d_book.m`
`plot_les_like.m`	2D LES-like analysis for `swnh2d_book.m`
`constant_N_seiche_freq.m`	Dispersion relation for constant N seiche
`constant_N_seiche_bookpics.m`	Particle transport by constant N seiche (pictures)
`constant_N_seiche.m`	Particle transport by constant N seiche (movie)
`constant_N_seiche_tracer.m`	Tracer transport by constant N seiche
`linear_lake.m`	Lake-like stratification and eigenfunctions
`variable_N_seiche_bookpics.m`	Particle transport variable $N(z)$
`variable_N_seiche_wbirthdeath.m`	Particles with birth and death variable $N(z)$
`simplest_particles.m`	Stokes problem with shear triggered swimming
`gyrotactic_stokes.m`	Stokes problem with gyrotactic swimming
`tracer1d.py`	1D tracer finite volume solver
`tracer2d.py`	2D tracer finite volume solver
`sw1d_fwave.py`	1D shallow water solver using the f-wave method
`sw_1d_nonhydro_fv.py`	Weakly non-hydrostatic 1D f-wave shallow water solver

Preface

While the aesthetic appreciation for the motion of fluids dates back to antiquity, the mathematical description of fluids in nature dates back to the mid-1700s when Leonard Euler posited the first equations of motion of a fluid as a continuum. This discovery followed from the discovery of the calculus of infinitesimals, which lends itself naturally to the "continuum approximation" – an approach that replaces finite fluid particles with a continuous space of matter at all length scales. As profound as this sounds (it is!), equations are nothing without solutions. Sadly, solutions resulting from classical mathematics, bound to "closed-form" and well-known functions such as sin, cos, exp, are as simplistic as they are esoteric. It would not be until the advent of numerical methods that progress towards "realism" in mathematical descriptions of fluids would be made. This is interesting because the complex, coherent, and vortical nature of natural fluid flow was clearly reflected in Leonardo da Vinci's sketches made hundreds of years before the invention of calculus.

In the early twentieth century Lewis Fry Richardson, in pursuit of the first numerical weather predictions, was the first to take Euler's description of fluids to task with the first "computers" – literally a room full of people carrying out numerical arithmetic calculations by hand! These early efforts were largely failures, but advances in technology after World War II soon spurred a massive increase in the range of fluid motions it became possible to meaningfully simulate. By the 1980s computational fluid dynamics had become solidly established as a scientific discipline.

The computers of the early 1980s were either room sized or desktops with very limited power. Much of this changed with the "PC revolution" that made significant computing power available even to non-specialists. Still, prior to the popularization of advanced graphical-user interfaces in the early 1990s, most computer experiences were text-based or relied on very simplistic graphical rendering (e.g., writing the color of a pixel to some predetermined memory block). The basic nature of the system resulted in rather crude and almost trivial user experiences, think e.g., Pong, Q*Bert, and Frogger. Yet, in spite of (or perhaps "because of") the crudeness, there was something "accessible" about it, in the sense that creating these early experiences was within reach of nearly anyone who could envision or dream of making the computer do something "just for them". The ability to compute for one's self is an incredibly empowering experience since it breaks one free from the shackles of somebody else's design and prescribed user experience patterns.

Interestingly, the quantitative description of population dynamics also has its roots in the 1700s, with the work of Malthus. The history of mathematics as applied to biology, like fluid mechanics, has its ebbs and flows, but its great flowering coincided with the availability of computation. The post WW II period, in particular, yielded massive advances, perhaps best exemplified for the purposes of this book by theory of pattern formation by Turing and the Hodgkin-Huxley model of the neuron as an excitable medium (both circa 1952). The minimal mathematical expression of the

xxvi Preface

former is as a set of coupled reaction diffusion equations, and of the latter model is as a set of nonlinear ordinary differential equations, one that is best treated using numerical solutions.

It is no surprise that the ability to execute simulations, store large amounts of information, and measure in novel, digital ways (e.g. satellite based remote sensing) serves as catalyst for scientific growth. The pace of this growth is so fast, it is almost impossible to survey by any one person, and indeed we are continually humbled when a reviewer points out a useful paper we had missed in some far flung corner of the literature. It has been pointed out many times, but bears repeating; few of us read the literature systematically often to the detriment of our work. Still, there is no feeling quite like formulating a simple model that captures the essence of a problem, coding it up and watching the results come to life in a visualization.

The codes in this book are not crude games with a limited artistic footprint, but rather are the solution to mathematical models posed as differential equations. Thus, these codes are not only bound by what is practically possible on the computing device, but also by what is theoretically possible for the equation sets themselves. The numerical methods thus represent equations in some discrete form on a digital computer. However, these two qualitatively different sets of constraints do not diminish the simulation enterprise, instead they make it richer and more rewarding. Physical realism is a worthy and deeply satisfying goal; think of how much better water is represented in today's animated films as opposed to those released in the early 2000s.

It is our hope that the novice computational scientist finds this book as an accessible means to gain a foothold into the sophisticated world of the numerical modeling of physics and ecology in natural waters. It is by no means extensive or complete, but it attempts to shed light on what is known at the time of writing, so readers can take their own ideas and simulations into the scientific arena.

Acknowledgments

MS would like to thank Michael Waite, Francis Poulin, and Kevin Lamb; his colleagues in the Environmental and Geophysical Fluid Mechanics group of the Applied Mathematics Department at the University of Waterloo for their collegiality and open sharing of ideas over his career. He would also like to thank Sue Ann Campbell, of the Mathematical Biology group of the Applied Mathematics Department at the University of Waterloo for many years of discussions of interesting mathematical neuroscience problems; in a highly siloed academic world it is a true pleasure to have colleagues who have time for those with a casual interest in fellow-traveler scientific disciplines.

The modeling point of view outlined in this book owes a huge debt to the many Continuum Mechanics (undergraduate) and Asymptotics and Perturbation Theory (graduate) students MS has taught at the University of Waterloo. It is a true honor to have interacted with this diverse and gifted group. Their influence on how I think is profound and far reaching, and much of it will be unconsciously reflected in these pages.

MS would also like to thank DS for friendship over the years. It is a rare relationship that covers professor–student, supervisor–student, and collaborator; rarer still when it is filled with as much humor and a deep commitment to loud music as ours has been.

DS would like to express his eternal gratitude to MS for taking a chance on a mediocre master's student in 2009 that ended up parlaying itself into a lifelong journey of learning, friendship, and pushing the boundary of the science and art that is numerical modeling. DS also thanks Kevin G. Lamb as well for the support throughout his Ph.D. studies and beyond, as his wisdom and ever-present scientific voice is always welcomed.

DS is also indebted to other his lifelong friends and applied math co-conspirators that helped squeeze enjoyment out of life at the best and worst of times. These include Michael Dunphy, Killian Miller, Kris Rowe, and Sumedh Joshi (who departed us all too early). DS also gratefully acknowledges the hosts of the 2012 Gene Golub Siam Summer school for teaching him the very valuable lessons he carries with him to this day and that continue to pay dividends, be they in work or academia: Randall LeVeque, Jörn Behrens, Frank Giraldo, and Michael Bader. It was a pivotal learning experience in his career who helped mold who he is today as a critical thinker.

CHAPTER

Introductory remarks and reader's guide

1

CONTENTS

1.1 Introduction.. 1
1.2 Reader's guide 5
References.. 7

1.1 Introduction

The motion of fluids, as a branch of physics has a history that dates back centuries. Yet the impact of technology on the subject is so profound, that the motion of natural waters can be considered to be a very young science. To put it plainly, the combination of remote sensing (especially satellite remote sensing) and ready access to computation has forever altered what was once a science of carefully acquired, widely separated in space observations, and theories largely based on the mathematics that was necessary to derive them. While for younger readers it may sound difficult to believe, even within one academic career, both authors can recall courses based entirely on perturbation expansions with ocean gyres described by theories completely devoid of the eddies we now know to be the *sine qua non* of the general ocean circulation! Sadly, while technology changes quickly, the University environment changes slowly. Courses maintain their structure long after the material taught in them is of ready importance to the next generation, and the general push to do more with less means facilities, be they experimental or computational, to teach larger classes from a modern point of view are often absent. Nevertheless, over the same academic career, the home department of one of us has gone from no mathematical biology courses to multiple mathematical biology courses at both the undergraduate and graduate levels. So we are left with remnants of the old, adrift in a sea of the new. Indeed, the books I learned from, associated with names like Gill ([1]), Whitham ([3]), and Murray ([2]) are still occasionally read by students, but just as often students use resources found by Google, or posted on YouTube.

One could rant against this loosening of the applied mathematics canon, but not only would this be pointless, it would also be unfair to the many excellent sources (not just books!) being published today. In fact, one could say that this is a Golden Age of academic publishing, with new resources and new ideas being produced at such a high rate it is almost impossible to keep up. So why another book? To put it succinctly: advection. Fluids flow, and natural waters are dominated by flow. The

Physics and Ecology in Fluids. https://doi.org/10.1016/B978-0-32-391244-0.00011-5
Copyright © 2023 Elsevier Inc. All rights reserved.

1

CHAPTER 1 Introductory remarks and reader's guide

mathematical description of flow is fundamentally nonlinear, and nonlinearity generally means fewer solutions in closed form (i.e. as formulae) and more analysis via numerical methods. At the same time, numerical methods, are often either self-taught (for those in the sciences) or taught as a branch of numerical analysis (for those in mathematics). Either way, young scientists are often left with misleading impressions of what is doable, even on a laptop, and lacking practice with the model-simulation-data analysis pipeline.

This book is an attempt to remedy this situation. It is not a classical applied mathematics book, though we do explain a fair bit of applied mathematics content using equations, as well as diagrams, analogies, and flow charts. It is also not a computer science book, though we do provide both code and discussions of algorithms and coding philosophy. It is a book about the dynamics of tracers in natural waters, where by "dynamics" we mean both the physics (simplified to some degree) and the biology (simplified to a much larger degree). Since science is a historically aware endeavor, some of what we discuss is older work that has its own mythology in certain circles (e.g. the theory of modes for lakes large enough to be affected by the Earth's rotation). This is often due to a fog of mathematical opacity, and we try to burn away this fog. Other aspects are new and may not even have a fully developed mathematical language. This too is a fact of life in modern science, though we do draw the line at purely data-centric approaches. We do so, not because we feel these have no value, but to contain the scope of the material to some degree.

Let us put the challenge we are posing to ourselves in context. Fig. 1.1 shows a satellite image, Creative Commons licensed and downloadable from

```
https://en.wikipedia.org/wiki/Algal_bloom
```

of the Western portion of lake Erie. The image is rich in detail, typical of the extraordinary potential of remote sensing. Several peninsulas protrude into the lake along (mainly) the Canadian (northern) shore. A Toxic algal bloom is visible in green. It takes the form of complex, yet coherent regions primarily visible near the northern shore. Being composed of miscroscopic algae, the blooms track the hydrodynamic motions in the lake, with a particularly prominent example being the vortex near the tip of the triangular Point Pelee. Moreover, while striking, the image in Fig. 1.1 is not atypical of the annual evolution of coexisting complex patterns biology and hydrodynamics in the lake.

The range of scales evident in the image, and the number of distinct types of phenomena in need of a mathematical description are not only impressive, but to anyone who has carried out mathematical modeling challenging enough to put a pit in the throat. Put simply there is a lot to do, with a lot of uncertainty in it!

The challenge of modeling the system is equaled by the importance of having accurate, reliable models. This is because the Lake Erie watershed is home to roughly 12 million people. As a result, Lake Erie experiences considerable water quality pressure and this is due to a combination of population pressure, industry, and farming run off. This has been documented in the literature since the 1970s. Despite of these challenges, the lake serves as a major source of recreational opportunities.

1.1 Introduction

FIGURE 1.1

Satellite image of a portion of Lake Erie during a toxic algae bloom. The bloom is prominent on the Canadian (northern) side of the lake and manifests in complex, yet coherent patterns. Note the clear vortex off the tip of the Point Pelee peninsula.

Due to the societal importance of the lake all sorts of groups of scientists turn their scientific attention on the lake. One particularly relevant example is that government agencies must, by law, provide predictions of certain quantities deemed of societal interest. Some of these involve multi-national agreements, while others provide the necessary information for municipal governments (e.g. should the municipal water pipes be closed due to a harmful algal bloom?). As part of the study of the lake simulations are carried out, often using complex, legacy models the inner workings of which are rarely examined in detail. Thus the majority of simulations do not follow the standard reductionist approach of theoretical physics. Indeed both the hydrodynamic and biological phenomena shown in Fig. 1.1 have active scientific communities questioning the "basics" of the models. This, of course, doesn't mean that nothing is known about them. However, some gray area exists as to what the "correct" model is for a particular level of detailed description. Correct is in quotes, because any model has to address the resources expended in using it (typically computer power and storage requirements).

CHAPTER 1 Introductory remarks and reader's guide

To put it another way, if one must prepare a water quality report every April to coincide with the end of the fiscal year, then there will be little time to pursue the "what if" questions of basic science. Yet Lake Erie is large enough to be affected by rotation, shallow enough so one might expect significant intrusions of near bottom turbulence into the main water column, biologically very active, with significant year to year variability in ice cover, all in a freshwater setting that has marked differences from the coastal ocean.

This book will not be about a comprehensive model for Lake Erie, or indeed any other lake. Instead, it will be all about first steps. Some will be modeling first steps, in which a quantity will be abstracted (think of population as mathematical function of time and space, as opposed to an actual population of individuals living in an actual lake) and then described by a conceptually simple relationship. A subset of these quantities will be physical, and their description will thus have several centuries of Newtonian mechanics to back them up. This will mean on the one hand that the reductionist approach to obtaining simpler models will be smoother, having a lot of past work to go on. On the other hand, the model reduction process will be expressed in the language of "classical" applied mathematics. This is problematic for the typical "science first" reader, because they have not had the long pre-requisite chains to build up the vocabulary that often clouds the description. Thankfully, numerical methods allow for a much easier approach to modeling. In effect, the computer becomes a virtual laboratory, allowing the scientist to pose scientific questions, that can be answered via graphs, or perhaps some mathematical analysis.

Other models discussed will be motivated by biology. The incredible diversity of life in natural waters means that the distance from reality to model is greater for biology than for physics. Thus what the reader will see will be extremely simplistic. Mathematical biology has grown by an incredible amount over the past 50 years, and we do not make any claims in superceding the many excellent books on the subject. Where we wish to make our mark, is in coupling biology with hydrodynamics; effectively putting advection in the driver's seat of our various explorations.

We are writing this book to effectively embolden readers. We want them to ask "what if" questions, to discover that even their desktop computer has amazing power, and that numerical methods can be relatively easy to use and modify.

Conceptually, the book is split into three inter-woven themes. The first theme is the narrative of how tracers are moved (advection), mixed (diffusion), and in the case of biological quantities, grow (reaction). We will build up a relatively complete description, falling back on conceptual descriptions and numerical experiments when the required mathematics would imply too large a digression (e.g. when we discuss the Navier Stokes equations we will not spend tens of pages delving into the concepts and manipulations of Cartesian tensors). The second theme aims to introduce various codes that solve and visualize mathematical problems we wish to discuss. In general the simple codes are presented in MATLAB®, while more complex codes are either in Python or C++. The codes are provided in well-documented form and this allows the reader to take any of a large number of useful building blocks and use them for their own explorations. They can also, if they wish, get an introduction into how scientific codes are written, and scaled up from the simple to the very complex. The third theme aims to make scientific observations based on codes and their results. This, in the authors' opinion, is the real art of modern scientific computing and we hope the readers get at least some of the fun we had in creating and using the codes. Throughout, we revisit a number of classical models, dispel some of the fog of complicated language around them, and demonstrate both how they work and how they fall short, in describing the dynamics in actual natural waters. We want the readers to come away emboldened, so that the frontiers of science continue to be pushed forward.

1.2 Reader's guide

While this book has been written to be read in sequential order, we recognized early in the writing process that the modern reader is often stretched for time. As a basic principle, if a reader has the choice between spending more time on the codes for a particular chapter, or quickly moving onto the next one, take the time to run and examine the codes. So much of what we describe in the pages of this book is experiential in nature, and after a couple of decades of our own modeling experience, we can safely say that the fun is in the results of the codes. This also means that there may be parts of the book where a reader may say "Hey, I know a better model than this". The modern scientific literature is vast, and the reader may well have found something we did not. We thus whole-heartedly support explorations that go beyond our own.

The graduate terms at the University of Waterloo where one of us teaches are 12 weeks long, and so the 12 chapters of the book could make for a one term course. If pressed for time, the more "show and tell" style chapters (1, 5, 12) could be assigned for reading, with the remainder kept for in class focus. The course would be well suited to a weekly classroom session (perhaps with slides) and a weekly computer lab session where some of the codes would be run in a group setting, with the chance for students to learn from each other, a TA, and the instructor. We have forgone traditional assignment problems, though do note when a calculation or detail may be a useful exercise of the reader. Instead we have provided a number of mini-projects for the reader. These alternate between uses for the codes provided with the book, and reading/research based projects. In a classroom setting, an instructor could use the provided codes to create short computer lab-based assignments, leaving the mini-projects for more substantial evaluations. We note that the ready access to platforms such as Zoom and Teams, coupled with the difficulty of fair online proctoring has led to a rediscovery of (typically short) oral exams. We believe the material presented in our book is ideal for such a form of evaluation.

It is also possible to use the book for self-study, either as whole, or with three different areas of focus. Fig. 1.2 shows the chapters for three shorter tracks through the book. Chapters relevant for a numerical methods focus are shown in blue, those relevant for a modeling focus are shown in green, while those relevant for a fluid mechanics focus are shown in orange. In our experience, many students in applied

CHAPTER 1 Introductory remarks and reader's guide

mathematics have courses with a numerical analysis focus available to them, but rarely do they have courses that aim to achieve multiple concrete goals numerically. Concrete goals, of course, require some context and this is why this track through the book material is the longest of the three presented. Part of what makes this book unique is our attempt to explain the core ideas of fluid mechanics without multiple courses of mathematical pre-requisites. As such Chapter 4 appears both in the modeling (green) and the fluid mechanics (orange) focused sequences. The ability to account for flow opens up a world to the modeler, and while we believe that there are many excellent books that focus on reaction and diffusion, it is the primary role of advection that we wish to convey to the reader.

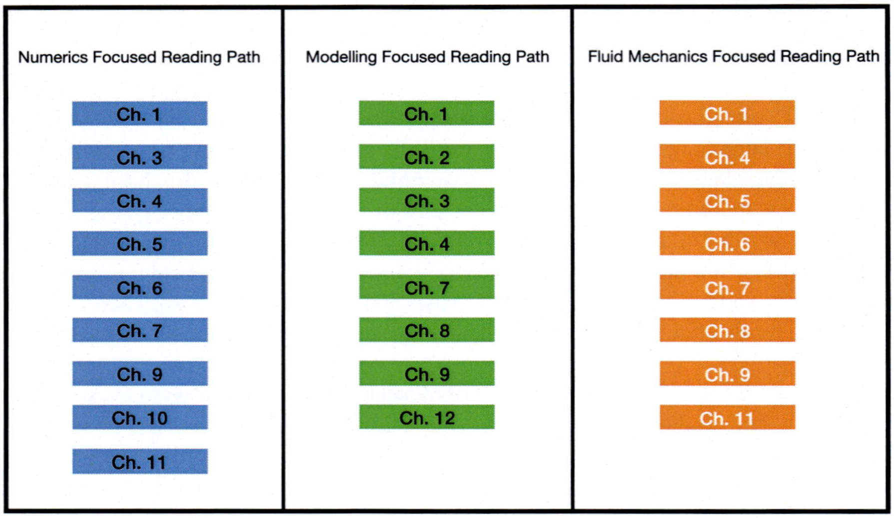

Numerics Focused Reading Path	Modelling Focused Reading Path	Fluid Mechanics Focused Reading Path
Ch. 1	Ch. 1	Ch. 1
Ch. 3	Ch. 2	Ch. 4
Ch. 4	Ch. 3	Ch. 5
Ch. 5	Ch. 4	Ch. 6
Ch. 6	Ch. 7	Ch. 7
Ch. 7	Ch. 8	Ch. 8
Ch. 9	Ch. 9	Ch. 9
Ch. 10	Ch. 12	Ch. 11
Ch. 11		

FIGURE 1.2

Three shorter tracks through the book. Chapters relevant for a numerical methods focus are shown in blue (in the left panel), those relevant for a modeling focus are shown in green (in the middle panel), while those relevant for a fluid mechanics focus are shown in orange (in the right panel).

The web page for the book

```
https://physicsandecologyinfluids.ca/
```

provides a link to relevant codes, and any errors or issues brought to the authors' attention via the community of readers.

Finally, we briefly comment on code, coding, and coding standards. As authors, we both come from an academic background that included a lot of code. This typically switched between high level languages like MATLAB and Maple, and low level languages such as Fortran and C, depending on the task that needed to be carried out. It also largely fell short of the professional coding standard of an ITworkplace (where one of us now spends their professional time). This does not mean that academic code is bad code, only that it is much less standardized. Python offers a modern, open source alternative for many applications, and at least potentially could bring academic coding closer to an industry standard. Indeed at the outset of this project, our plan was to create a unified software framework using Python as a wrapper for a low level language package. The disruption of the Covid-19 pandemic led us to reconsider. In the end, code is meant to work, and in the applied mathematical setting, to carry out mathematical tasks and show their results graphically. Once the minimum standard of commenting is reached, there are no further style points! Thus many of the codes that are distributed as part of this book are in MATLAB. As such, they have been more thoroughly tested by various undergraduate and graduate students. Readers are welcome to translate to any language they see fit. Of course, when a code sooner or later swells to above a few hundred lines it becomes more important to break it into multiple files, and begin focusing on a more industry-like approach where more modern coding practices such as encapsulation and code re-use principles are used. Open-source projects that one shares with colleagues, or intends to publish to the web, are encouraged to adopt and adhere to their own consistent stylistic guidelines on e.g., white-space use, consistent formatting/indentation, and descriptive naming conventions so they might appeal to a broader audience. In such "open" settings, code adopt-ability and extensibility hinges more on readability and usability aspects than in one's personal numerical experimentation laboratory.

While many of the codes for this book are written in a digestible format in short, self-contained MATLAB or Python scripts, the scripts relying on the Discontinuous Galerkin Finite Element Method (DG-FEM) (see Chapters 10 and 11) depend on the C++ library entitled 'blitzdg' written by one of the authors. This is due to the fact that the high-order DG-FEM method relies on heavy mathematical machinery such as orthogonal polynomials, unstructured triangular grids, and linear algebra routines that would essentially be boilerplate if included in the scripts. The library's python wrapper pyblitzdg is meant to provide an easy-to-install and simple interface to these more sophisticated mathematical tools, and the details of blitzdg are left in its GitHub repository for those who are interested enough to go and find it. Those who are not, can usually install the library and its wrapper in their own local python environment by issuing the following terminal command:

```
pip install pyblitzdg
```

since the package is available publicly on the python package index: https://pypi.org.

References

[1] A. Gill, Atmosphere-Ocean Dynamics, 1st ed., Academic Press, 1982.
[2] James Dickson Murray, Mathematical Biology: I. An Introduction, Springer, 2002.
[3] G.B. Whitham, Linear and Nonlinear Waves, Wiley-Interscience, 1999.

CHAPTER

The basics of modeling: Mechanics and population models

2

CONTENTS

2.1 A (not so) basic model... 9
2.2 Exponential growth 13
2.3 Logistic growth.. 14
2.4 The Lotka-Volterra model 15
2.5 Extending the Lotka-Volterra model 18
2.6 Models of plankton populations.. 25
2.7 Mini-projects 27
2.8 Computing spectra 28
References.. 30

2.1 A (not so) basic model

Applied mathematics is full of "basic" models. These have, over the past centuries, led to a wealth of theory, usually related to the fact that the resulting differential equations could not be solved in closed form. Often a student is thrown into the middle of one of these developments without a firm handle on the assumptions behind the model. Herein, we want to start with, what will be in the end, a very simple model: namely the damped simple harmonic oscillator (often abbreviated as the SHO). However, we want to talk a little bit about how we get to the model. In Fig. 2.1 we show the basic definition diagram for the mass on a spring model. There are a couple of key assumptions, though neither is thought of as terribly controversial:

1. The mass can be treated as a point mass
2. The spring force is proportional to the change in the spring's length and acts to bring the spring to its undisturbed length

Notice these were much longer to write out in words than their mathematical expression. The first, after all, basically means we can write the **position** of the mass as $x(t)$ without which we couldn't even contemplate an differential equation-based model, while the second just means $F(t) = -kx(t)$ where we assume the mass is originally located at $x = 0$.

Physics and Ecology in Fluids. https://doi.org/10.1016/B978-0-32-391244-0.00012-7
Copyright © 2023 Elsevier Inc. All rights reserved.

10 CHAPTER 2 The basics of modeling

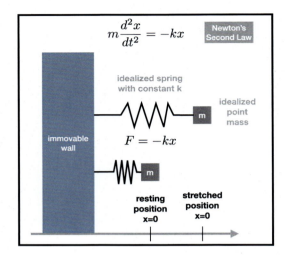

FIGURE 2.1

The definition diagram for the mass on a spring as an example of the simple harmonic oscillator or SHO.

The resulting ordinary differential equation (henceforth ODE), a restatement of Newton's Second Law,

$$m\frac{d^2x}{dt^2} = -kx \tag{2.1}$$

can be written in what is called the simple harmonic oscillator (or SHO for short) form

$$\frac{d^2x}{dt^2} + \omega_0^2 x = 0 \tag{2.2}$$

where ω_0 is the natural frequency of the system. For the mass on the spring we have

$$\omega_0 = \sqrt{\frac{k}{m}}. \tag{2.3}$$

If we recall that the period is $T = 2\pi/\omega_0$, this rather simple expression has some interesting implications, namely that the natural period of oscillation for a mass-spring system increases with mass and decreases with the spring constant. Sometimes we say the spring is "harder" as k gets larger, since a fixed displacement would mean a larger force. It is also interesting that the period depends on the square roots of k and m, as opposed to just their values. In practice this means a small change in spring constant is likely to only affect the period a tiny bit. This means oscillators are rather forgiving, and helps explain how first year physics laboratories subjected to the wear and tear of generations of first year classes can give reasonable results year after year!

Something more complex happens if the mass is very small, but this is beyond the present discussion.

Mathematically the SHO can be solved rather trivially,

$$x(t) = A\cos(\omega_0 t) + B\sin(\omega_0 t) \tag{2.4}$$

and this solution can tell us just about anything we may wish to know about the motion. However, because exact solutions are more the exception rather than the rule, and because quantities like energy are important both physically and mathematically, we will derive the energy equation. Recall that the kinetic energy (KE) is given by

$$E_k(t) = \frac{1}{2}mv^2 = \frac{1}{2}m\left[\frac{dx}{dt}\right]^2 \tag{2.5}$$

If we take the time derivative and use the Chain Rule we find that

$$\frac{dE_k}{dt} = m\frac{dx}{dt}\frac{d^2x}{dt^2}.$$

Next we use the governing equation (2.1) to find that

$$\frac{dE_k}{dt} = \frac{dx}{dt}[-kx(t)]$$

which does not seem to have gotten us very far. However if we look back at how the Chain Rule worked for $\frac{dE_k}{dt}$ a bit of trial and error will show that

$$\frac{d}{dt}\left[\frac{1}{2}kx(t)^2\right] = kx(t)\frac{dx}{dt}.$$

So we define the spring energy; really an example of potential energy (PE); as

$$U(t) = \frac{1}{2}kx^2 \tag{2.6}$$

and now the rate of change of kinetic energy is easily found to be

$$\frac{dE_k}{dt} = -\frac{dU}{dt}.$$

This is a bit non-traditional from a physics point of view, but actually kind of useful since it tells us that a loss of KE is exactly matched by a gain in PE and vice versa. The more traditional way to write it is

$$\frac{d}{dt}[E_k(t) + U(t)] = 0 \tag{2.7}$$

which states that the total energy of the system is constant.

12 **CHAPTER 2** The basics of modeling

Mathematics is, in many ways, a giant game of "How can I generalize that fact?" So let's see what aspects of the above model can be generalized. Let's start with what shouldn't be changed, and that is the definition of kinetic energy. It is true that first year physics is full of 'leaking sandbag' problems and other ways to mess with Newton's second law, but once the differential equation point of view on mechanics is fair game, there is not really all that much to it, one just changes $F = ma$ to 'the sum of forces equals the rate of change of linear momentum with time'. It is likely you will see a problem of that sort on an assignment. The true low hanging fruit lies in the spring energy term, because it was identified based on an analogy with taking the derivative of the kinetic energy and applying the chain rule. So what if we said that in general

$$U(t) = U(x(t)) \tag{2.8}$$

and tried to get an idea of what the resulting restoring force would be and whether this made sense for as broad of a class of U as we could reasonably say something for. Well the force part is pretty easy. For our example

$$U = \frac{1}{2}kx^2$$

and

$$F = -kx$$

so that we see right away

$$F = -\frac{dU}{dx} = -U'(x).$$

Those of you with a background in physics will recognize this definition (in higher dimensions the force vector is proportional to the gradient of the potential, but we won't need that fact here). The more interesting question is: What sorts of $U(x)$ are possible? And perhaps more importantly, what other choices than a quadratic are useful? To answer this recall the basic fact that the restoring force should move the mass back to its undisturbed location. This means when $x > 0$ we have $F < 0$ and for $x < 0$ we have $F > 0$. For $U(x) = ax^n$ we see right away that we need a positive and n even. Thus for example we could take

$$U(x) = \frac{1}{2}kx^2 + \frac{1}{4}k_3x^4$$

so that the restoring force is

$$F(x) = -kx - k_3x^3.$$

For small x the new, cubic piece is not that important and for larger x it means the restoring force is larger than it would be for the linear spring. Sometimes this is called a 'hardening spring'. The energy equation for such a spring states that

$$\frac{1}{2}mv(t)^2 + \frac{1}{2}kx(t)^2 + \frac{1}{4}k_3x(t)^4 = \text{constant}$$

but the resulting ODE

$$m\frac{d^2x}{dt^2} + kx + k_3x^3 = 0$$

is nonlinear and generally not taught in undergraduate ODE courses. For those interested, it is often called Duffing's equation, see the nice Wikipedia page dedicated to it for more info:

```
https://en.wikipedia.org/wiki/Duffing_equation
```

The previous discussion provides us with enough to go on as we move forward, but from the point of view of "realism" it falls short since we have not discussed the dissipation of mechanical energy into heat. There are a number of good reference books for this (Taylor's book [5] being a particularly excellent choice) and we leave the interested reader to pursue these.

2.2 Exponential growth

Let's start with some basic mathematical abstraction. Our independent variable will be time, t and our dependent variable will be the **population** $P(t)$. We will assume population is real valued and that time various continuously. Obviously 35.6 individuals is a pretty silly way to measure population, so we have to understand that we should be thinking about large values of population so that the error we incur by counting fractions of individuals is negligible. We are also not going to worry about spatial distribution of individuals in the population. We will include a description of spatial distribution in later chapters. The simplest model of birth and death asserts that the rate of change of population with time, due to both birth and death, is proportional to the present population. If α represents the ratio of the population that gives birth and β the ratio of the population that dies off, the fundamental differential equation governing population growth is

$$\frac{dP}{dt} = \alpha P - \beta P. \tag{2.9}$$

The initial value problem (labeled IVP by mathematicians) with $P(0) = P_0$ can then be solved using standard first year calculus to yield the exponential population law

$$P(t) = P_0 \exp[(\alpha - \beta)t]. \tag{2.10}$$

The essence of this model is that whether the population grows in an unbounded way, or decays to zero, is completely determined by the sign of $\alpha - \beta$, or in other words

14 **CHAPTER 2** The basics of modeling

whether the birth rate exceeds the death rate. When the birth rate exceeds the death rate, the model exhibits the pathology that the population increases without an upper bound. When the death rate exceeds the birth rate the model has the somewhat more subtle pathology that the population tends toward zero but never quite gets there. Put another way, the population never goes extinct!

2.3 Logistic growth

The most basic critique of the exponential growth model is that it allows for unbounded growth, and any population eventually reaches a point where its consumption of resources exceeds the availability of resources. One possible response is to say that the exponential growth model is not complete enough to cover this situation and should be re-examined as far as validity is concerned when the population is large. Notice, this isn't a mathematical criticism, but a modeling one. The math can be perfect, but the model could be useless. A more measured response may ask what the simplest fix for finite resources could be? To answer this we must introduce a second abstraction, namely the notion of **carrying capacity**. The carrying capacity, K is defined as the maximum population that an environment can stably sustain. If we let $r = \alpha - \beta$, and assume $r > 0$ so that population grows, the logistic population model has the governing differential equation

$$\frac{dP}{dt} = rP\left(1 - \frac{P}{K}\right). \tag{2.11}$$

In introductory mathematics courses on differential equations students are given this equation and asked to show that $P = K$ is a stable equilibrium. This is a nice exercise, but in many ways it is completely backwards, since the definition of carrying capacity is such that it has to be a stable equilibrium. Indeed if the model said it wasn't, we would have to change the model! This is a very important point; the machinery of mathematics is huge, useful and quite possibly overwhelming in terms of language and notation, but it is the application (hopefully through data) that determines if a model is useful. It is also possible to solve the IVP for (2.11) exactly (take $P(0) = P_0$). One useful way to write the solution is to let

$$q = \frac{K - P_0}{P_0}$$

and write the solution as

$$P(t) = \frac{K}{1 + q \exp(-rt)}. \tag{2.12}$$

Notice that the above tells us that whether P_0 starts below or above K is immaterial to the long time behavior. In all cases we tend to K exponentially fast.

2.4 The Lotka-Volterra model

We have discussed the dynamics of a single population, for example in contrasting exponential and logistic growth. However most populations interact with other living (e.g. predators) and non-living (e.g. nutrients) things. In this section we present the simplest predator-prey model, often referred to as the Lotka-Volterra model. We assume that the DE for the prey population (label it $C(t)$) has an exponential growth term as well as a death, or predation, term which is proportional to both the population of prey and population of predator (which we label $F(t)$ after the "fox" of the typical rabbit-fox example of a predator-prey). The predator is assumed to die off at a constant rate, but to have a birth rate which is proportional to both the population of prey and population of predator. The complete equations can be written as

$$\frac{dC}{dt} = \alpha C - \beta C F \tag{2.13}$$

$$\frac{dF}{dt} = \delta C F - \xi F. \tag{2.14}$$

All of the constants are assumed to be positive. In terms of their meaning, α is the net natural growth rate for the prey population, while ξ is the net natural death rate for the predator population. This is a key point: the predator cannot survive in this model without consuming prey. The model is thus not really viable without the nonlinear terms. The rate of consumption of prey is proportional to both the predator and prey populations, with a rate constant β. Since a predator cannot efficiently consume all of a prey a further constant, $0 < \delta < \beta$ specifies how efficient or inefficient this process is.

If we do the standard thing from differential equation theory and look for equilibria or steady states, we readily find that $(C, F) = (0, 0)$ is one. It takes a bit of algebra to find a second, but in the end one finds

$$(C, F) = \left(\frac{\xi}{\delta}, \frac{\alpha}{\beta} \right)$$

The interested reader can then follow the methodology of stability of equilibria, linearize around these steady states, and perform the eigenvalue analysis to check for stability as an *Exercise*. Here we are mostly concerned with the qualitative aspects of the model and for that a numerical solution is sufficient. The code used to generate the figures is

```
lotka_volterra_plain.m
```

and can be modified by the user as they see fit. Four sample time series are shown in Fig. 2.2 for four different initial conditions for the prey ($C(0) = (0.1, 0.5, 5, 10)$). Only the second half of the total integration time is shown. The time series demonstrates the well-known periodic cycle of predator and prey. The prey population (in black) peaks first, and this is followed by a rapid increase in predators (in blue). This

16 CHAPTER 2 The basics of modeling

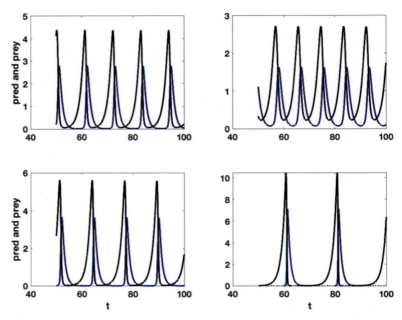

FIGURE 2.2

$C(t)$ (black) and $F(t)$ (blue) versus t for the Lotka-Volterra model. Only the second half of the integration time shown. $C(0) = (0.1, 0.5, 5, 10)$ for the four panels with $C(0) = 0.1$ in the top left and $C(0) = 10$ in the bottom right.

leads to a rapid decrease in prey, leaving nothing for predators to eat, and then the cycle repeats. The nonlinearity of the system is reflected in the profound asymmetry of the cycle. In particular, the predators have a sharp increase and a much slower decrease. This is often labeled as a **relaxation oscillation**. It is clear that while periodic, the solutions for C and F are very far from the sinusoids that are taught in calculus courses as exemplars of periodic functions.

The corresponding phase plane plots, showing the shape of an orbit on C and F axes, are shown in Fig. 2.3.

The phase space plot illustrates a rather poignant criticism of the Lotka-Volterra model, namely the tiny values of the populations of both the predator and prey during the "down" portions of the cycle. There are many different avenues for improving the Lotka-Volterra system, but for our purposes we will accept the model as a useful, if imperfect tool. To conclude, let's scale out, or non-dimensionalize the problem. This can be approached on formal mathematical grounds, or strictly as utilitarian: we would like a model with the smallest number of parameters possible. There is one independent variable, t, and two dependent variables, C and F. We can define dimensionless values as

$$(C, F, t) = (X\tilde{C}, Y\tilde{F}, T\tilde{t})$$

2.4 The Lotka-Volterra model

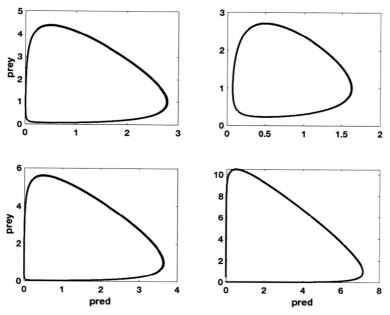

FIGURE 2.3

Phase space plot for the Lotka-Volterra model. Only the second half of the integration time shown. $C(0) = (0.1, 0.5, 5, 10)$ for the four panels with $C(0) = 0.1$ in the top left and $C(0) = 10$ in the bottom right.

so that the governing system of ODEs can be written as

$$\frac{X}{T}\frac{\tilde{C}(\tilde{t})}{d\tilde{t}} = \alpha X \tilde{C} - \beta XY \tilde{C} \tilde{F}$$
$$\frac{Y}{T}\frac{\tilde{F}(\tilde{t})}{d\tilde{t}}(\tilde{t}) = \delta XY \tilde{C} \tilde{F} - \xi Y \tilde{C}, \qquad (2.15)$$

or on simplifying and dropping tildes

$$\frac{dC}{dt} = \alpha TC - \beta TYCF$$
$$\frac{dF}{dt} = \delta XTCF - \xi TF. \qquad (2.16)$$

If we let $T = 1/\xi$, $a = \alpha/\xi$, $Y = \xi/\beta$ and $X = \xi/\delta$, we get the simpler system

$$\frac{dC}{dt} = aC - CF \qquad (2.17)$$
$$\frac{dF}{dt} = CF - F. \qquad (2.18)$$

18 CHAPTER 2 The basics of modeling

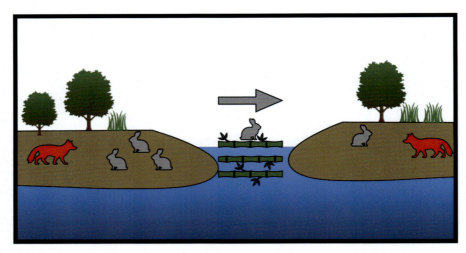

FIGURE 2.4

Diagram of the two island Lotka-Volterra model with only rabbits being able to cross between islands. Rabbits move from the island with higher rabbit population to the one with a lower population.

Of course the initial conditions must be scaled as well, but from Fig. 2.3 we can see that the initial conditions are more or less irrelevant, since the solution tends to the periodic solution, or in more mathematical words, to a **stable limit cycle**. Thus, if we were asked to characterize the solution behavior for a broad range of parameters, instead of having to vary four parameters we just have to vary one (a). There are many different avenues for improving the Lotka-Volterra system, which we will explore in the following section. In its basic form the Lotka-Volterra has an obvious analogy to the SHO of mechanics, with the important difference that the nonlinearity (i.e. the predation term) is essential.

2.5 Extending the Lotka-Volterra model

A central aspect of what we wish to impart on the reader throughout this book is that numerical methods allow for the exploration of numerical experiments in direct analogy of what one might do in a physical laboratory. In some sense, this removes the shackles associated with mathematical theory, because we are no longer dependent on exact solutions (i.e. formulae). Of course, mathematical theory is not removed from the picture entirely, because mathematical vocabulary and ideas are how the numerical experiments are described and analyzed.

In this section we explore two extensions of the Lotka-Volterra model from the previous section. The first of these is schematized in Fig. 2.4. The model describes

2.5 Extending the Lotka-Volterra model

the populations of foxes (predators) and rabbits (prey) on two islands. It is assumed that neither animal can swim, but rabbits can cross between islands over a "bridge" that is not strong enough for foxes to cross (schematized by floating bamboo in the figure). Rabbits cross from the island with a higher population to the one with a lower population. In the interest of having the simplest model the rate of crossing is assumed to be proportional to the difference of populations, namely if the population of rabbits on island one is larger than on island two, $C_1 > C_2$ then the new term in the island one rabbit population equation reads

$$- k(C_1 - C_2). \tag{2.19}$$

The complete model thus reads

$$\frac{dC_1}{dt} = \alpha_1 C_1 - \beta_1 C_1 F_1 - k(C_1 - C_2) \tag{2.20}$$

$$\frac{dF_1}{dt} = \delta_1 C_1 F_1 - \xi F_1. \tag{2.21}$$

$$\frac{dC_2}{dt} = \alpha_2 C_2 - \beta_2 C_2 F_2 - k(C_2 - C_1) \tag{2.22}$$

$$\frac{dF_2}{dt} = \delta_2 C_2 F_2 - \xi F_2. \tag{2.23}$$

It is important to note that if we were to add the two rabbit population equations, the term expressing crossing between the islands would cancel out. The model has considerable freedom in terms of parameters, since we have allowed for completely different parameters on the two islands. In practice, we would need to be constrained by ecological considerations.

The equations are solved via a standard second order Runge-Kutta scheme. For the first experiment we change the birth rate of rabbits between the two islands so that island 2 has a birth rate that is 89% of that on island 1. We then vary the rate at which rabbits cross between islands, showing the four values

$$k = (0, 0.0001, 0.001, 0.01)$$

in the two figures we produce. The code used to generate the figures is

```
lotka_volterra_two.m
```

Fig. 2.5 shows the phase portraits for the four values of inter-island exchange of rabbits. The top left panel is what numerical experimentalists would call a "sanity check"; since we set the rate of rabbits crossing between islands to zero. This panel shows that in this case we indeed get two very similar looking limit cycles for the two islands. The difference in shape is due to the small change in rabbit birth rate, but the qualitative behavior is the same. The bottom right panel shows the most extreme case of the strongest rabbit exchange between islands. Note the change of scale from the top left figure. It is clear that the evolution slowly spirals toward a single point, meaning that the two islands both tend to a stable steady state. That is a clear qualitative

20 CHAPTER 2 The basics of modeling

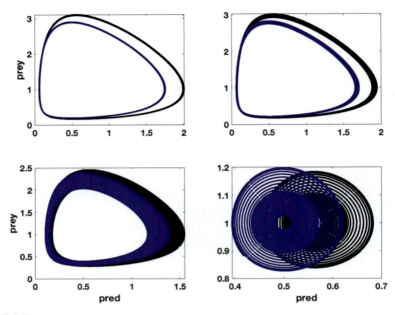

FIGURE 2.5

Phase portraits for the two island model, experiment one (change in rabbit birth rate). $k = (0, 0.0001, 0.001, 0.01)$ for the four panels with $k = 0$ in the top left. Island 1 is shown in black, island 2 in blue.

difference in behavior and as such a very interesting finding! It is also one that would be quite difficult to establish without numerical experimentation.

The top right and bottom left panels show us what happens for weaker between-island rabbit exchange. The top left panel appears to show the limit cycles found without exchange, drawn with a thick marker. This is because there has been a shift away from the stable limit cycle without between island rabbit exchange, to orbits that decay very slowly to what is presumably a stable equilibrium. As the exchange rate is increased the rate of decay increases (bottom left panel).

The numerical experiment has identified a clear mathematical difference between the case without exchange and cases with exchange. As such, if we were after a mathematical result, we could pursue the rigorous proof that a positive k implies a stable equilibrium for the four populations. Scientifically, we are in somewhat murkier waters. A four ODE model is clearly a cartoon of the real world, and as such assumptions such as the constant birth rate/death rate may not be appropriate. As such behavior in the top left and top right panels might be very difficult to distinguish in real world measurements.

This observation is given more credence if one computes the **spectra** of the prey populations shown in Fig. 2.6. Spectra are effectively the breakdown of the time series of the solution into their sinusoidal constituents and form a standard tool of data

2.5 Extending the Lotka-Volterra model

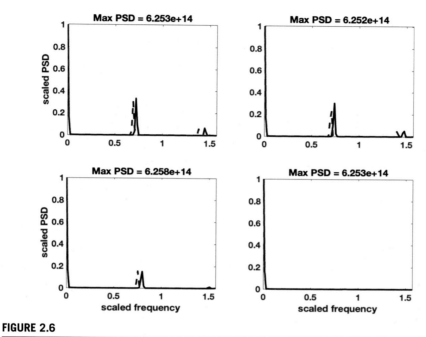

FIGURE 2.6

Spectra (the Power Spectral Density or PSD) for the two island model, experiment one (change in rabbit birth rate). $k = (0, 0.0001, 0.001, 0.01)$ for the four panels with $k = 0$ in the top left. Island 1 is shown in black, island 2 in blue.

analysis. If the notion of spectra is new to the reader, see the final section of this chapter for an elementary introduction. The quantity shown is called the **Power Spectral Density** or (PSD). For the present example, it can be seen that the two populations have a clear period of oscillation, with the top left (no exchange) and top right (weak exchange) cases showing a similar secondary peak for higher frequencies (faster oscillations). Thus from the point of view of spectral analysis, we would be likely to conclude that these two cases are very similar indeed. The bottom row shows that as the exchange between the islands is strengthened, the spectra do change. In the bottom left the secondary peak vanishes, and the primary peak decreases in magnitude by almost 50% compared to the top right panel. In the bottom right the spectrum is dominated by a peak at a zero frequency, confirming that we are at, or near, a steady state.

This relatively simple numerical experiment clearly outlined a difference in philosophy. Numerical experimentation is exploratory in nature, and is best when cases are inter-compared (in as organized a way as possible). Rarely are the black or white conclusions of mathematics reached, though in some cases mathematical work (i.e. rigorous proofs) can be motivated by numerical experimentation.

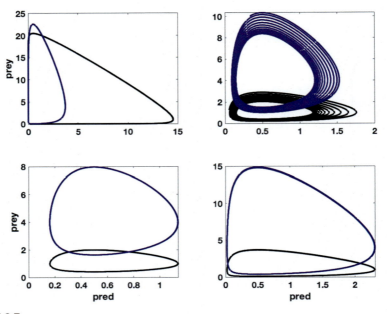

FIGURE 2.7

Phase portraits for the two island model, experiment two (change in predation rate). $k = (0, 0.0001, 0.001, 0.01)$ for the four panels with $k = 0$ in the top left. Island 1 is shown in black, island 2 in blue.

While our first experiment could be deemed a "success" the fact is that we made a change of a single parameter in a nonlinear model (the rabbit birth rate). The question one can ask is "how generic is the behavior observed"? This is often a thorny issue in mathematical modeling, and the present model is no exception. In the absence of field or experimental data, a series of trial and error explorations led to an instructive second experiment. For the second experiment we fix all Lotka-Volterra parameters save for the predation efficiency parameter δ which we decrease by 75% for island 2. The code used to generate the figures is

```
lotka_volterra_twob.m
```

Fig. 2.7 again shows the phase portraits as the rate of between-island exchange of rabbits increase from $k = 0$ in the top left panel, to $k = 0.01$ in the bottom right panel. The top left panel establishes the differences that changing predation rate induces. As expected the number of predators on island two is reduced compared to island one, because predators are less efficient at extracting a reproductive advantage from predation. As the inter-island exchange increases the observed behavior is somewhat surprising. In the top right panel, with a very weak exchange of prey $k = 0.0001$, it appears that the limit cycles are replaced by a slow decay to stable equilibrium populations on both island. The same comments made above for experiment 1 apply

2.5 Extending the Lotka-Volterra model   23

here; the importance of the mathematical distinction between a limit cycle and a stable equilibrium would have to be assessed by comparing the time scale of decay to the time scales on which we might expect model parameters to change (e.g. with the changing of seasons). Putting these aside for the moment, we can see as the rabbit exchange rate between islands is increased further in the bottom left and bottom right panels, the system returns to limit cycle behavior. Examining the extent of the axes one notes that the range of predator population is quite small, compared to the top left case with $k = 0$. Nevertheless, we appear to have recovered a different parameter regime in which exchange between islands is an essential part of setting the properties of the limit cycles observed.

We have hinted above that the various constants of the Lotka-Volterra model may, in fact, not be constants at all. From an increased death rate during a harsh winter, to a consistent breeding season that yields the births of a population there are multiple natural cycles that may be essential to include in the modeling exercise. Our second model considers a relatively simple introduction of time dependence into the Lotka-Volterra model via a periodic change of the prey birth rate. The governing equations read,

$$\frac{dC}{dt} = \alpha(t)C - \beta CF \tag{2.24}$$

$$\frac{dF}{dt} = \delta CF - \xi F \tag{2.25}$$

$$\alpha(t) = \alpha_0(1 + a\sin(2\pi t/T)). \tag{2.26}$$

Here a is a fractional measure of how much the birth rate changes, and T sets the period over which the birth rate increases and decreases. For readers familiar with the way in which ordinary differential equations are generally taught, this system, through the explicit dependence on time on the right hand side, immediately places the model outside of the scope of many standard theorems of differential equation theory. From a simulation point of view, the model is no more difficult to integrate than the constant coefficient model discussed above.

For the numerical experiment we wish to discuss, we set $a = 0.4$ and $T = 20$. These values were chosen by experimentation with the code. Different suggestions are discussed in the Exploratory Projects section below. We follow the same general structure as the two experiments with the two island model, except instead of varying the coupling ratio, we vary the initial condition for the prey choosing the same as we did for the basic Lotka-Volterra model. The code used to generate the figures is

```
lotka_volterra_time_dep.m
```

Fig. 2.8 shows the phase portraits as the initial condition is varied. There is a qualitative difference between the top row (small initial rabbit populations) and the bottom row (large initial rabbit populations). If we contrast this with what was shown in Fig. 2.3 we see that the phase portraits in the lower row of Fig. 2.8 have a larger range of variation, but maintain some of the features of the bottom left panel of Fig. 2.3.

24 CHAPTER 2 The basics of modeling

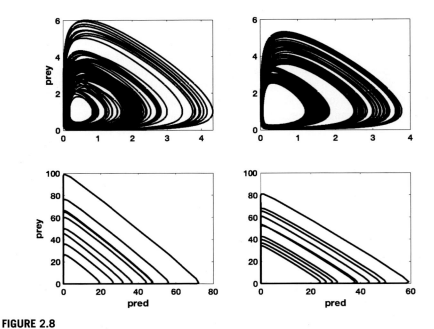

FIGURE 2.8

Phase portraits for the time dependent birth rate model ($a = 0.4$, $T = 20$). $C(0) = (0.1, 0.5, 5, 10)$ for the four panels with $C(0) = 0.1$ in the top left and $C(0) = 10$ in the bottom right.

The lower values of initial prey population are even more interesting, because the time dependent birth rate case in the top row of Fig. 2.8 has far more complex orbits. Just how complex some of these orbitsare can be quantified with spectral analysis.

Fig. 2.9 shows the spectra for the same four cases as Fig. 2.8. The frequencies have been normalized by the frequency with which the birth rate changes. If the system were linear, the expected response would thus be found near a frequency of 1. The values of the power spectral density have been scaled by the maximum of the case shown in the top left panel. In each case the spectrum is dominated values at zero frequency, indicating a non-zero mean value. We thus restrict the y axis between 0 and 0.5. The spectra show some unexpected features. First of all for low initial values of prey concentration, the response has 3–5 peaks. For the top left panel these do not occur at multiples of the frequency at which the birth rate varies, but for the top right panel a dominant response at twice the frequency of the variation of the birth rate is observed. Thus while the top row of Fig. 2.8 appears quite irregular, the spectra suggest it can be thought of as the combination of a few frequencies. The bottom row of Fig. 2.9 shows a marked contrast, with many frequencies having significant power. This is sometimes termed a broad band response and is often associated with the onset of chaos (for more on chaos see [4]). It is impressive with a relatively simple change

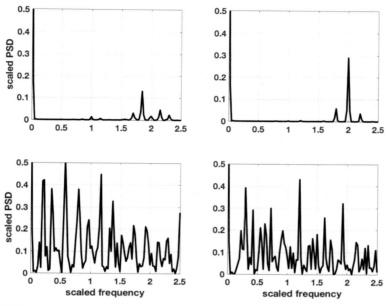

FIGURE 2.9

Phase portraits for the time dependent birth rate model ($a = 0.4$, $T = 20$). $C(0) = (0.1, 0.5, 5, 10)$ for the four panels with $C(0) = 0.1$ in the top left and $C(0) = 10$ in the bottom right.

to the basic Lotka-Volterra model (and without the goal of inducing such behavior) we were able to get such a complex response.

2.6 Models of plankton populations

Plankton is a catchall term for a large family of microscopic living things that live in natural waters. Traditionally plankton is divided into animal-like zooplankton, and plant-like phytoplankton. Biologists who study plankton have much finer distinctions than this (for example, see the approachable book [2] for a thorough discussion of phytoplankton, of course, but in the spirit of the simplest possible model let us consider the bulk populations of zooplankton and phytoplankton to be denoted by $Z(t)$ and $P(t)$, respectively.

In what is now a classical paper by Truscott and Brindley ([6]) the phytoplankton population is modeled as a logistic growth law, with an additive term for predation by zooplankton. This term is like that in the Lotka-Volterra model, except it is thought vital to include a rate limitation as the phytoplankton population gets large. The gov-

26 **CHAPTER 2** The basics of modeling

erning equation reads

$$\frac{dP}{dt} = rP\left(1 - \frac{P}{K}\right) - R_m Z \frac{P^2}{\alpha^2 + P^2}. \tag{2.27}$$

The four parameters represent: the inherent growth rate (r), the predator-free carrying capacity (K), the maximum specific predation rate (R_m) and the rate at which the maximum predation is achieved as a function of phytoplankton population α.

The zooplankton is assumed to only be able to increase its population due to predation, so that

$$\frac{dZ}{dt} = \gamma R_m Z \frac{P^2}{\alpha^2 + P^2} - \mu Z. \tag{2.28}$$

Here μ is the death rate of zooplankton, and γ is somewhat crude device for representing how efficiently predation is converted into increasing the population of zooplankton.

Truscott and Brindley give the typical parameter values as

$$K = 108 \ \mu\text{g N/l}$$
$$r = 0.31/\text{day}$$
$$R_m = 0.71/\text{day}$$
$$\alpha = 5.7 \ \mu\text{g N/l}$$
$$\mu = 0.0121/\text{day}$$
$$\gamma = 0.05.$$

Of note among these is the low death rate of zooplankton and the rather inefficient conversion of predation into population increase for the zooplankton. It is also worth noting that for many plankton models the time scale is that of days. Since most fluid mechanics models use the MKS system of units, this means care must be taken to ensure any coupled models have conversion factors to ensure a consistent treatment of time.

Truscott and Brindley perform a fairly standard dynamical systems analysis of the equations (see [4] for an introduction), and provide numerical evidence for "bursting" or red tide behavior. In the spirit of this chapter we show sample solutions that explore parameter space. The code used to generate the figures is

```
truscott_brindley_plain.m
```

Fig. 2.10 shows numerical integrations of the Truscott Brindley model for four different initial phytoplankton populations. The lower left panel reproduces Truscott and Brindley's Figure 4. The "bursting" which makes this model so well known is evident for the lower row, but low initial values of phytoplankton population lead to a steady state, as opposed to bursting.

Of course the Truscott and Brindley model has been superseded by many far more complex models. Many of these have a strong and independent literature ([3]) that a

2.7 Mini-projects

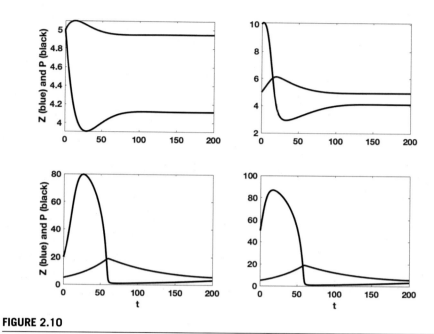

FIGURE 2.10

Evolution of the Truscott Brindley model from four different initial phytoplankton populations. The bottom left panel reproduces their Figure 4.

brief survey would not do justice to. Simple models like Truscott and Brindley ([6]) do have a role in coupled fluid dynamics-plankton models and this is a topic we will revisit in future chapters.

2.7 Mini-projects

Mathematics books are typically judged by the quality of the problem sets provided. In a work whose purpose is to encourage numerical exploration we will mostly avoid textbook problems, choosing instead to present little mini-projects that can be explored with small modifications of the codes provided. Below are five mini-projects related to the material presented above, all of which can be carried out with small changes to the codes provided.

1. Modify the

    lotka_volterra_plain.m

    code to show pairs of trajectories for two initially close by values of predator and prey population. How does the distance between trajectories change with time?

28   **CHAPTER 2** The basics of modeling

2. Modify the

    lotka_volterra_time_dep.m

code so that the death rate increases due to harsh winters.
3. Modify the

    lotka_volterra_two.m

code so that only the predators move between islands. Reproduce the existing experiments and compare and contrast with the results in the text.
4. Modify the time dependent Lotka-Volterra model so that the predators hibernate (they do not die, hunt or give birth during this time period). Modify the

    lotka_volterra_time_dep.m

code for this case.
5. Modify the plain Lotka-Volterra model so that the there are two groups of rabbits. Young rabbits and mature rabbits. Assume the predators preferentially hunt young rabbit at double the rate of mature rabbits. Modify the

    lotka_volterra_plain.m

code for this case and explore its phase space behavior.

2.8 Computing spectra

The simple harmonic oscillator gives solutions that are sinusoidal. Indeed many problems in classical mechanics give solutions that are some combination of sinusoids. Famous examples include then normal modes of a double pendulum, the motion of a mass on a strain hardening spring (i.e. a Duffing oscillator). See [5] for examples and details of the classical mechanics. On the mathematical side, the Fourier series provides an almost magical way to approximate periodic, or periodically extended functions. Given this compact, and efficient manner with which Fourier series represent functions, it should be no surprise that Fourier type methods have been used for data analysis, and especially time series. In fact in many engineering disciplines a Fourier based course in time series methods would be considered standard. Here we will pursue a somewhat ahistorical, though hopefully intuitive, approach.

As is the case of many codes, the numerical method used to solve the two island Lotka-Volterra model gives values for the various variables at discrete times which are regularly spaced. We will use a subscript to denote the discrete value, and denote the variable as $x(t)$ for illustrative purposes. We are thus working with x_n where $n = 1, \cdots, N$. For now, we will assume that we chose N so that $x_1 = 0$ and x_N is the point right before $x = 0$ again. In terms of the sine and cosine basis of Fourier series you can think of this as not wanting to repeat the point of exact periodicity at the very end. This turns out to not be a strong assumption, and we discuss this below.

By taking the ends to be essentially zero, we are able to think of using a Fourier sine series. This isn't strictly necessary, but for an illustrative discussion it helps to focus on a simpler basis.

Next we have to answer the twin questions of what is the longest and shortest period sine we can think of describing with our data. Because our data is equally spaced in time (this is important), we know the time extent of the time series to be $N \Delta t$. Thus the longest period sine we can represent is precisely one copy of the sine, or one with a period

$$T_{max} = N \Delta t.$$

What about the shortest possible period? Well here we have to first agree that we aren't saying that we hope to represent this short period sine terribly well. Instead we want to do the mathematical thing and say how short would the period have to be so we can't possible represent it? One way to think about that problem is to say with three adjacent times we should be able to represent the three zeros of a single period of sine. If we fix the left endpoint to be $\tilde{t} = 0$, then the first zero occurs at $\tilde{t} = \Delta t$ and the second at $2 \Delta t$. Thus any sine with a period less than $2 \Delta t$ cannot have a full period represented by the grid. Thus

$$T_{min} = 2 \Delta t.$$

In terms of frequency we say that the smallest possible frequency we can represent is

$$\omega_{min} = \frac{2\pi}{T_{max}} = \frac{2\pi}{N \Delta t}$$

Moreover we can represent integer multiples of this frequency up to

$$\omega_{max} = \frac{2\pi}{T_{min}} = \frac{\pi}{\Delta t}.$$

The representation of the time series we have shown is thus a form of the Fourier series

$$x_k = \sum_{n=1}^{N} b_n \sin(\omega_n k \Delta t).$$

The coefficient b_n could be found by numerical integration, but one of the reasons why the theory of time series runs so deep in the engineering community, is that the computation can be done much faster using the famous Fast Fourier Transform, or FFT (see the excellent discussion in [1]). Interestingly, the FFT goes back to Gauss, and hence predates modern computation by a considerable amount of time. Regardless of how we do it, once we have b_n we can define the **spectrum** of our time series (or sometimes the **power spectral density**, or **PSD**) at each frequency as

$$S(\omega_n) = b_n^2.$$

CHAPTER 2 The basics of modeling

A signal can be thought of as "simple" when b_n is significant for only a few n. On the other hand a signal with many significant values of b_n can be thought of as complex, or even one obtained from a chaotic process.

The relation between the spectrum and the total "energy" of the signal is given by the rather famous Parseval's theorem, so that

$$\sum_{n=1}^{N} b_n^2 = \sum_{n=1}^{N} x_n^2.$$

While the above is fairly clear, with implementation details provided in the MATLAB® code, we caution the reader that the general theory of Fourier analysis of time series data deserves an independent reading.

There is one bit of notation that is worth pointing out. In the time series literature it is usually written that

$$\omega_n = n \frac{\pi}{\Delta t} \text{ where } n = -\frac{N}{2}, \cdots, \frac{N}{2}$$

so that when $n = N/2$ we recover ω_{max}. In any event the really high frequency content in a time series is rarely of much interest.

Above we mentioned that it isn't a very strong assumption that the time series is periodic, with one end at zero, and the other very close to it. This may have seemed to the reader like some pretty serious chutzpah on our part. The reason for our statement is that in practice, time series are "windowed". While the theory is beyond our elementary theoretical discussion, the idea is that if we take an arbitrary function $x(t)$ and multiply it by a window function $w(t)$ that is close to one over the majority of the domain, and zero at the ends (with a nice, smooth decay) we are guaranteed to have a well-behaved time series. Many software packages (e.g. MATLAB, scipy) will have ready implementations of any number of windowing methods. For the interested reader we suggest [1] for further exploration.

References

[1] Brad G. Osgood, Lectures on the Fourier Transform and Its Applications, vol. 33, American Mathematical Soc., 2019.

[2] Colin S. Reynolds, The Ecology of Phytoplankton, Cambridge University Press, 2006.

[3] Karline Soetaert, Peter M.J. Herman, A Practical Guide to Ecological Modelling: Using R as a Simulation Platform, vol. 7, Springer, 2009.

[4] Steven H. Strogatz, Nonlinear Dynamics and Chaos: With Applications to Physics, Biology, Chemistry, and Engineering, CRC Press, 2018.

[5] John R. Taylor, Classical Mechanics, University Science Books, 2005.

[6] J.E. Truscott, J. Brindley, Ocean plankton populations as excitable media, Bulletin of Mathematical Biology 56 (5) (1994) 981–998.

CHAPTER

Modeling active tracers

3

CONTENTS

3.1 Numerical experiments 31
3.2 Active tracers: the simplest models 32
3.3 Nonlinear effects 34
3.4 Interacting tracers: the Lotka-Volterra model.................. 36
3.5 Advection-diffusion for plankton models 39
3.6 Tracers in two dimensions 42
3.7 Advection diffusion population model in 2D 46
3.8 Mini-projects 49
References... 50

3.1 Numerical experiments

In the previous chapter we have developed the basics of modeling by introducing the simple harmonic oscillator of classical mechanics, and a series of gradually more complex population models. These ranged from models that could be solved in terms of formulae, to those that needed to be solved numerically. We used the second, more complex class of models, to introduce the framework of "numerical experiments", in which the computer is used as a virtual laboratory. In this chapter we seek to build up an understanding of 'dynamics' from the point of view of populations, or concentrations of tracer, that are distributed in space. In the language of applied mathematics we will consider both passive and active tracers, and we will consider models of advection, reaction, and diffusion. Many existing mathematical biology sources concentrate on reaction diffusion (e.g. [3]), so it is really advection that is the major novelty in what we present. The spirit of the presentation is to build on the few closed form solutions, but in the end, concentrate less on these and more on illustrative computational experiments. For readers looking for particular topics the chapter summary is

1. Linear models of a single tracer: Population evolution, population with advection and population with diffusion
2. Nonlinear extensions
3. Tracers in two-dimensions
4. Populations in two-dimensions

Physics and Ecology in Fluids. https://doi.org/10.1016/B978-0-32-391244-0.00013-9
Copyright © 2023 Elsevier Inc. All rights reserved.

32 **CHAPTER 3** Modeling active tracers

3.2 Active tracers: the simplest models

While physicists have models so well established (e.g. plane waves) that their utility is almost never questioned, the same is not true for other sciences. In the case of biology, this comes down to the deep, philosophical question of whether reductionism is the appropriate philosophy to use when attempting to gain an understanding of a biological system (e.g. a food web in a lake with excess inflow of fertilizer). We will accept the basic notion of reductionism (i.e. splitting up a complex phenomenon into simpler parts) in what follows, and hence take a concentration of some material $C(x, t)$ as our basic variable of interest. If we were to take C to represent a population, we saw in the previous chapter that the simplest model for population growth or decay is the exponential model

$$\frac{dC}{dt} = \alpha C \tag{3.1}$$

where the sign of α tells us whether the population grows ($\alpha > 0$) or decays ($\alpha < 0$). This model has no spatial information, and from the various classical partial differential equations two come to mind as simple, readily interpretable extensions of the above model. The first example adds advection by a known, constant current, $u_0 > 0$. The so-called reaction advection equation in this case reads,

$$\frac{\partial C}{\partial t} = -u_0 \frac{\partial C}{\partial x} + \alpha C. \tag{3.2}$$

The solution of this equation can be derived by various mathematical means, but the intuition behind the solution does not require any mathematics at all. Consider a system governed by the exponential law (3.1) and now place it on a conveyor belt moving at a constant speed u_0. The growth and decay is not affected by the motion and the solution will have the form

$$C(x, t) = C(0) \exp(\alpha(x - u_0 t)). \tag{3.3}$$

The combination of variables $x - u_0 t$ is simply a way to account for the left to right motion of the conveyor belt. This is quite a useful result, because it tells us that for a more complex system the "reaction" part of the system proceeds largely independently of the advection. Thus a numerical code can concentrate on the "best" technique for each of the parts and worry about coupling between the two processes to a lesser degree.

The second simple model worth considering combines the reaction term with a diffusion term, to produce a reaction-diffusion equation. Such equations are broadly studied in mathematical biology, and we only seek the briefest of glimpses at the theory. The simplest reaction diffusion equation reads

$$\frac{\partial C}{\partial t} = D \frac{\partial^2 C}{\partial x^2} + \alpha C. \tag{3.4}$$

3.2 Active tracers: the simplest models 33

Those who have been exposed to the theory of PDEs will know that while diffusion equations are well behaved mathematically, they do not have a particularly large number of exact solutions. We can, however borrow an idea from the physics of waves and try to guess at a solution that oscillates in space:

$$C(x, t) = \cos(kx)c(t).$$

Substitution gives the ordinary differential equation for lower case $c(t)$

$$\frac{dc}{dt} = (-Dk^2 + \alpha)c,$$

which is an exponential form equation with a growth/decay constant that depends on the biophysical parameters α and D as well as the length scale of the spatial oscillation, k. The solution is

$$C(x, t) = c(0)\cos(kx)\exp([\alpha - k^2 D]t), \qquad (3.5)$$

and tells us that even if $\alpha > 0$ so that a population 'naturally' grows, the effect of diffusion can lead to decay. Since k is the wavenumber, hence inversely proportional to wavelength, shorter waves will be damped more. This is quite a find, since it is an explicit demonstration of how a spatial process (diffusion) can overcome a temporal one (exponential growth).

If we put it all together we get a model that is in some sense the most complex possible; it includes exponential growth or decay, diffusion and advection,

$$\frac{\partial C}{\partial t} = -u_0 \frac{\partial C}{\partial x} + D \frac{\partial^2 C}{\partial x^2} + \alpha C. \qquad (3.6)$$

At the same time it is also the simplest such model possible since we are ignoring nonlinear effects, chose a growth law that is essentially trivial and allow only one spatial dimension. Nevertheless, the result does not have a simple, closed form solution in physical space. Instead we solve the system numerically using a Fourier spectral method (i.e. by Fourier transforming we can solve in Fourier space exactly and hence only use the Fast Fourier Transform (henceforth **FFT**) to transform to physical space). The script is called

`adv_diff_react_1d.m`

and a sample solution produced by this script is shown in Fig. 3.1. The initial state consists of one wavepacket (in the right half of the plot) and one wavepacket envelope (in the left half of the plot). It can be seen that while the envelope propagates and grows in time, the packet decays due to the effect of diffusion.

Wave evolution processes are often visualized by a special type of plot, called the space-time or Hovmoeller plot. An example of this type of plot is shown in Fig. 3.2. In a plot of this type, with the time dimension represented by the y-axis, a rightward tilt indicates a signal that is propagating to the right. It can be seen that the wide

34 CHAPTER 3 Modeling active tracers

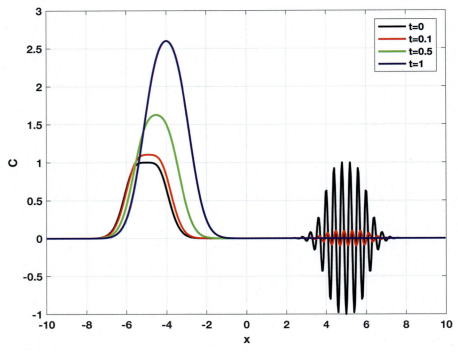

FIGURE 3.1

The evolution of an advection, reaction diffusion equation. The initial state is shown in black. Later times (in dimensionless units) are shown in red, green, and blue for $t = 0.1, 0.5$, and 1.

packet envelope initially centered at $x = -5$ clearly propagates to the right, spreads a bit due to diffusion and increases in amplitude. The wave packet initially centered at $x = 5$, in contrast is invisible after about $t = 0.2$. Interestingly, before vanishing, the packet is seen to propagate a small distance to the right (evidenced by the small tilt to the right in the Hovmoeller plot).

3.3 Nonlinear effects

There are two obvious nonlinear effects we can imagine incorporating into the models discussed in the previous section. The first is a quadratic nonlinear term of the form found in the material derivative (see the discussion in Chapter 4)

$$C\frac{\partial C}{\partial x}.$$

3.3 Nonlinear effects

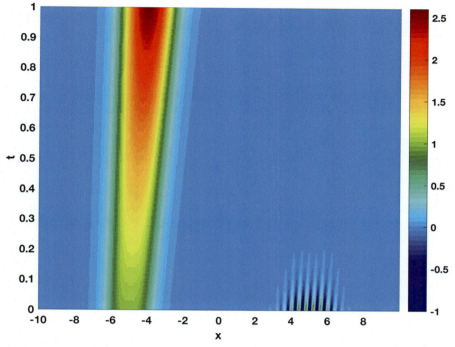

FIGURE 3.2

The evolution of an advection, reaction diffusion equation shown as a space time plot.

The second is a more realistic growth law, in its most general form $G(C)$, where $G(C) = \alpha C$ is the exponential case discussed previously. While many $G(C)$ have been discussed in the literature, perhaps the most famous extension of the exponential growth law is that of logistic growth

$$G(C) = \alpha C \left(1 - \frac{C}{C_{carrying}}\right) \quad (3.7)$$

where $C_{carrying}$ is called the carrying capacity of the system. The so-called logistic equation

$$\frac{dC}{dt} = \alpha C \left(1 - \frac{C}{C_{carrying}}\right)$$

yields exponential growth when C is small, but growth slows near $C_{carrying}$, the carrying capacity of the system. When $C > C_{carrying}$ the rate of change of C with time is negative, so that while $C = 0$ is an unstable steady state, $C = C_{carrying}$ is a stable steady state.

36 **CHAPTER 3** Modeling active tracers

The combination of diffusion and logistic growth leads to the well-studied Fisher's equation (sometimes known as the Fisher-KPP equation),

$$\frac{\partial C}{\partial t} = \nu \frac{\partial^2 C}{\partial x^2} + \alpha C(1 - C) \qquad (3.8)$$

where we have scaled the concentration by the carrying capacity. While the question of whether a solution can be expressed in closed form is quite complicated, traveling wave solutions can be sought by introducing a traveling variable $y = x - Vt$ so that Fisher's equation reduces to

$$-VC' = \nu C'' + \alpha C(1 - C)$$

and this can in turn be converted to a system of two first order ordinary differential equations. Using standard dynamical systems theory it is possible to prove that a family of traveling wave solutions exists ([2]). It is perhaps easier to solve Fisher's equation numerically. The script used to solve this is equation is

`adv_diff_fisher.m`

and an example solution is provided in Fig. 3.3 (the script allows the user to generate more complex plots). It can be seen that the combination of diffusion and reaction leads to a propagating front. This is not a wave in the sense of physics, since Fisher's equation has no representation of a restoring force. One can think of it as a reaction front, or more colorfully as a fire that burns a region, but having done so, cannot do it again.

A combination of quadratic nonlinearity and diffusion is also a "named" equation; the Burgers equation

$$C_t + CC_x = \nu C_{xx}. \qquad (3.9)$$

As an aside, since the equation is named after Johannes Burgers there is no apostrophe in the equation's name. The Burgers equation has a storied past in applied mathematics, since it can be transformed to the linear diffusion equation through the well-known Cole-Hopf transformation (see the classical book [4] for more information). Moreover Like Fisher's equation, the Burgers equation also has a traveling front solution, though for the applications we are interested in the Fisher's equation is perhaps the better conceptual model.

3.4 Interacting tracers: the Lotka-Volterra model

OK, now let's reconsider the Lotka-Volterra model with populations that vary in space as well as time. This type of model can be written in a standard form as a system

$$\frac{\partial C}{\partial t} = -u_0 \frac{\partial C}{\partial x} + \nu \frac{\partial^2 C}{\partial x^2} + G_1(C, F)$$

3.4 Interacting tracers: the Lotka-Volterra model

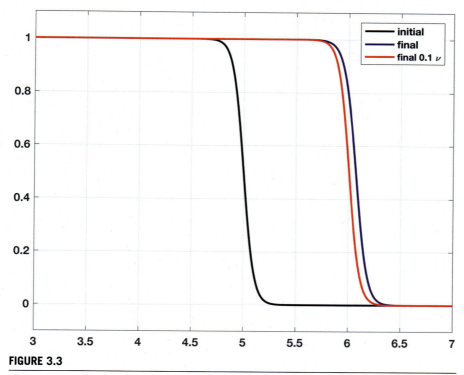

FIGURE 3.3

The evolution of two examples of Fisher's equation. The initial state is shown in black, with a later time in blue and red (the latter for ν reduced by an order of magnitude).

$$\frac{\partial F}{\partial t} = -u_0 \frac{\partial F}{\partial x} + \nu \frac{\partial^2 F}{\partial x^2} + G_2(C, F). \quad (3.10)$$

The left hand side is the rate of change in time and on the right hand side, from left to right, the terms are labeled as advection, diffusion and reaction (interaction might be a better catchall, but reaction is widely used). For the scaled Lotka-Volterra model we have

$$[G_1, G_2] = (aC - CF, CF - F).$$

This type of model is quite easy to solve numerically if we are not worried about modeling near-boundary behavior. For a domain with periodic boundary conditions, the spatial derivatives can be handled by the Fourier transform (the FFT in practice) and the reaction terms can be handled by any number of ODE solvers (e.g. the Runge-Kutta family of solvers).

Before showing some numerical solutions, consider first the case with pure advection ($\nu = 0$). If we define a new, moving spatial variable $y = x - u_0 t$, and write $[C, F] = [C(y, t), F(y, t)]$ we can see that the advection terms cancel and we are left

CHAPTER 3 Modeling active tracers

with just the ODE-based, classical Lotka-Volterra system. This result, of course, is quite sensible, because a population of foxes and rabbits on Earth is, as a point of fact, hurtling through space at a rather fast speed. Put another way, putting a population onto a conveyor belt should not change the dynamics of that population.

Adding diffusion complicates the picture, because in addition to the predation interaction between the two populations, diffusion means that populations interact with their neighbors (i.e. individuals move from areas with high population to areas with low population). We saw this in the previous chapter for the two island ordinary differential equation model ((2.20), (2.21), (2.22), (2.23)).

For the numerical simulations we choose to perturb the population of the prey in a systematic manner using a combination of a wavepacket envelope and a shorter length scale wavepacket:

$$C(x, 0) = 0.3 + 0.4 \text{sech} \left(\frac{x+5}{1} \right)^2 + \text{sech} \left(\frac{x-5}{1} \right)^2 \cos \left(\frac{2\pi x}{0.4} \right)$$

while the population of predator is perturbed with a draw from a uniform distribution. In MATLAB® this is accomplished by

$$F = 0.05 + 0.1 \, \text{rand}(\text{size}(C)).$$

We fix $a = 0.2$ and $u_0 = 0.25$. The questions we would like to answer are:

1. Is the limit cycle pattern maintained?
2. What is the effect of the initial perturbations?

To accomplish this we use the code in

```
adv_diff_lotka_volterra.m
```

which has more can reproduce the figures shown in the text, but also has additional functionality in terms of plots for the reader to experiment with. In Fig. 3.4 we show a case with low diffusion, $v = 10^{-3}$. It can be seen immediately from the red-blue band structure in the vertical that the periodic, or in other words limit cycle, behavior of the ODE Lotka Volterra system is maintained. We can see that the effect of the initial perturbations is visible as a signal that 'tilts to the right' indicating propagation. Interestingly spatial variations are maintained for a considerable amount of time, and somewhat surprisingly, the short length scale packet leads to a strong perturbation even at late times (note the bump near $x = -6$ for $30 < t < 40$). In a more complex model, one should thus be careful about attributing any prominent 'bump' in a signal purely locally. Indeed once an event of interest is identified, a careful analysis should attempt to trace the event back in time to its spatiotemporal point of origin.

In Fig. 3.5 we show a case with higher diffusion, $v = 10^{-2}$. While the general features described in the previous paragraph are evident they are far more smoothed out, and more difficult to identify. This is an important cautionary tale, because many basin scale models have both 'modeled' turbulent diffusion and implicit diffusion based on the numerical scheme. Fig. 3.5 clearly shows that controlling excess diffusion matters for accurate modeling of biophysical systems.

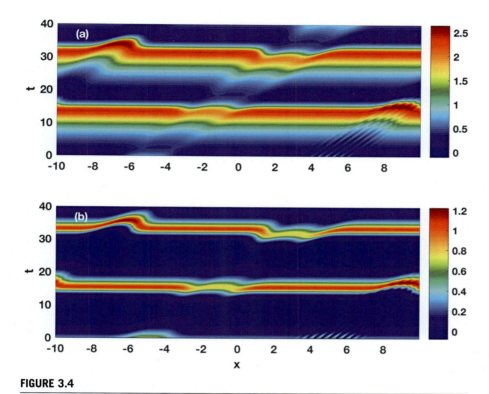

FIGURE 3.4

Space-time, or Hovmoeller plots of C (a) and F (b) with $\nu = 10^{-3}$. Note the way in which the initial population perturbation signal slopes to the right for both the broad packet envelope and the shorter length scale packet. Red-blue bands indicate the periodic, or limit cycle behavior in time.

3.5 Advection-diffusion for plankton models

We discussed the zooplankton-phytoplankton (ZP) model due to Turcott and Brindley in the previous chapter (Eqs. (2.27), (2.28)). We showed numerically that it does indeed lead to excitability, or a spiking behavior over some part of parameter space. Here we reconsider the model as a one dimensional advection-reaction-diffusion model:

$$\frac{\partial P}{\partial t} = -u_0 \frac{\partial P}{\partial t} + \kappa \frac{\partial^2 P}{\partial x^2} + rP\left(1 - \frac{P}{K}\right) - R_m Z \frac{P^2}{\alpha^2 + P^2} \quad (3.11)$$

$$\frac{\partial Z}{\partial t} = -u_0 \frac{\partial Z}{\partial t} + \kappa \frac{\partial^2 Z}{\partial x^2} + \gamma R_m Z \frac{P^2}{\alpha^2 + P^2} - \mu Z. \quad (3.12)$$

CHAPTER 3 Modeling active tracers

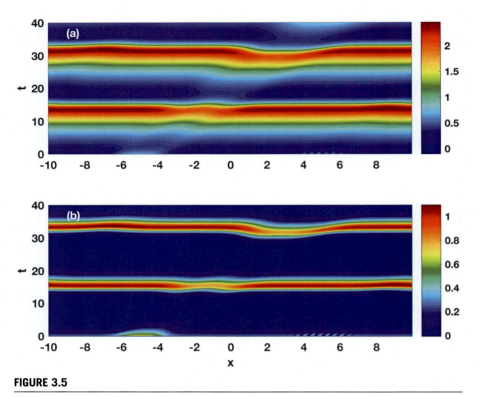

FIGURE 3.5

Space-time, or Hovmoeller plots of C (a) and F (b) with $\nu = 10^{-2}$. Note the way in which the initial population perturbation signal slopes to the right for both the broad packet envelope and the shorter length scale packet. Red-blue bands indicate the periodic, or limit cycle behavior in time.

Recall that for the parameter set we used in the previous chapter the relevant time scale was in days. We reuse the same parameters, and choose the two remaining parameters u_0 and κ as parameters of convenience. The code used to generate the figures is

adv_diff_tandb.m

We initialize Z near $Z = 5$ with uniformly distributed random noise between zero and two added. In contrast the $P(t = 0)$ variable is specified according to a deterministic distribution

$$P(x, 0) = 15 + 20\text{sech}\left(\frac{x+5}{1}\right)^2 + \text{sech}\left(\frac{(x-5)}{1}\right)^2 \cos\left(\frac{2\pi x}{0.4}\right).$$

3.5 Advection-diffusion for plankton models

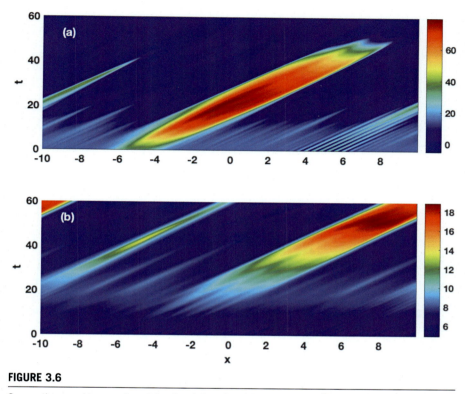

FIGURE 3.6

Space-time, or Hovmoeller plots of P (a) and Z (b) with $\nu = 10^{-3}$. Note the way in which the initial population perturbation signal slopes to the right for both the broad packet envelope for P. Propagation in the Z field is only evident after $t = 20$.

While this is largely chosen for illustrative convenience (like the previous case both a large perturbation and a modulated wavepacket perturbation are included), it is in part based on the sensitivity to initial conditions study in the past chapter.

Fig. 3.6 shows Hovmoeller plots for the P and Z fields. It can be seen that for the P variable both the large perturbation and the packet-like perturbation lead to propagation (i.e. rightward sloping regions). The wave packet like perturbation is smoothed by viscosity until short scale fluctuations are no longer evident after around $t = 20$. In contrast the Z field takes some time to grow, with propagation only clearly evident until after $t = 20$. In Fig. 3.7 the same information is repeated but as color coded line plots with t increasing by 10 units between different colors. The line plots are particularly striking in showing how different the late time and early time evolution of the Z field are.

The reader is encouraged to try different initial condition values to explore how the system behaves. One particularly interesting effect is observable when the diffusivity is lowered (say to 10^{-4}).

42 CHAPTER 3 Modeling active tracers

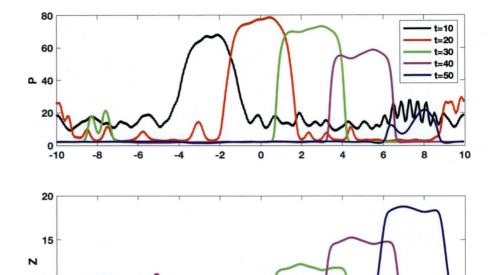

FIGURE 3.7

Line plots of P (a) and Z (b) with $\nu = 10^{-3}$. t increases from $t = 10$ by 10 time units.

3.6 Tracers in two dimensions

In the previous section we discussed the possible increase in complexity when multiple tracers, or multiple populations are considered. In this section we discuss the increase in complexity due to flow fields in two dimensions. In two dimensions the equation governing a tracer that is advected by a known velocity field $\vec{u}(x, y) = (u(x, y), v(x, y))$ reads

$$\frac{\partial C}{\partial x} = -\frac{\partial u C}{\partial x} - \frac{\partial v C}{\partial y} + \nu \left(\frac{\partial^2 C}{\partial x^2} + \frac{\partial^2 C}{\partial y^2} \right) \tag{3.13}$$

or in vector calculus form

$$\frac{\partial C}{\partial t} = -\nabla \cdot (C\vec{u}) + \nu \nabla \cdot \nabla C.$$

This form is useful if one integrates over a region and applies Gauss' Theorem,

$$\frac{d}{dt} \iint_D C \, dA = -\int_{\partial D} C\vec{u} \cdot \hat{n} \, dS + \nu \int_{\partial D} \nabla C \cdot \hat{n} \, dS.$$

3.6 Tracers in two dimensions 43

Here the left hand side represents the rate of change of total tracer in the region D, while the right hand side represents the two fluxes across the boundary:

1. The advective flux $\vec{J}_{adv} C \vec{u} \cdot \hat{n}$
2. The diffusive flux $\vec{J}_{diff} = -\nu \nabla C \cdot \hat{n}$

The unit normal, $\hat{n}$, is assumed to point outward.

In a true model of a basin the velocity field is specified by its own governing equations, but for now let's consider a few simple instructive fields we can specify. Let's consider a simple shear flow $\vec{u} = (y, 0)$ so that the governing equation simplifies to

$$\frac{\partial C}{\partial t} = -y \frac{\partial C}{\partial x} + \nu \left(\frac{\partial^2 C}{\partial x^2} + \frac{\partial^2 C}{\partial y^2} \right). \tag{3.14}$$

When the diffusion is neglected we get an even simpler equation, namely

$$\frac{\partial C}{\partial t} = -y \frac{\partial C}{\partial x}.$$

In this equation the y coordinate is just a parameter and the D'Alembert solution of the advection can be used to write down a solution to the initial value problem with $C(x, y, t = 0) = F(x)$

$$C(x, y, t) = F(x - yt). \tag{3.15}$$

It is easy to see from this solution for $y > 0$ the solution is advected to the right, while for $y < 0$ the solution is advected to the left. Shear thus spreads tracer and diffusion is not needed to do so.

When diffusion is not negligible the problem is best handled numerically. Since boundaries are not important, we again use FFT based methods. These are implemented in the code

adv_diff_shear.m

which allows the user to generate a number of different plots and to change the model parameters.

Fig. 3.8 shows the concentration field at time 0 (panel (a)) and at a later time $t = 3$ for two values of the diffusion parameter (the larger value in panel (c)). It can be seen that the shear flow acts to spread the tracer, leading to significantly more diffusion. This enhancement of spreading by shear is called "shear dispersion" or "Taylor dispersion" (after G.I. Taylor, an eminent 20th century fluid mechanician) in the literature. The shear creates thin regions which the diffusion then acts to smooth out. The contrast between panel (b) and (c) shows how weaker diffusion maintains the 'stirring' of the tracer by the shear flow.

Of course a shear flow is only the simplest possible case. A type of flow that allows us to explore the key role of diffusion in the evolution of tracers, and especially

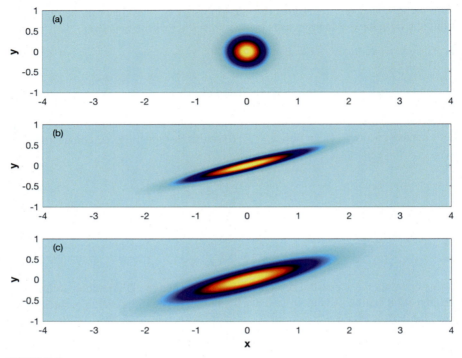

FIGURE 3.8

Snapshots of the concentration for the advection by a shear field with diffusion equation. (a) $C(x, y, t = 0)$, (b) $C(x, y, t = 3)$ with $\nu = 10^{-3}$, (c) $C(x, y, t = 3)$ with $\nu = 10^{-2}$.

active tracers, is one that has clear boundaries that a non-diffusive tracer cannot cross. In the theory of fluid mechanics, incompressible flows in two dimensions (or those for which $\nabla \cdot \vec{u} = 0$) can be derived from a stream function $\psi(x, y, t)$ so that the flow is along streamlines ψ = constant (see [1], or Chapter 8). This means if we have regions with closed streamlines a tracer initially found in these regions cannot cross streamlines and "get out". For concreteness we choose the streamfunction

$$\psi = \sin(\pi x) \cos(\pi y/2) \tag{3.16}$$

so that

$$u = \frac{\partial \psi}{\partial y} = -\frac{\pi}{2} \sin(\pi x) \sin(\pi y) \tag{3.17}$$

and

$$v = -\frac{\partial \psi}{\partial x} = -\pi \cos(\pi x) \cos(\pi y/2). \tag{3.18}$$

3.6 Tracers in two dimensions 45

We fine tune the initial distribution of tracer a little bit from the usual Gaussian "bump" profile. By doing so, we create a smoothed square by using higher, even powers in the exponential

$$C(x, y, t = 0) = \exp\left[-\frac{((x - x_0)^n + (y - y_0)^n)}{w_d^n}\right], \qquad (3.19)$$

where the parameters we choose are

$$x_0 = 0.5$$
$$y_0 = -0.85$$
$$n = 8$$
$$w_d = 0.1$$

meaning that the distribution of tracer is a square in the rightmost circulation cell that is subsequently advected and diffused.

Using the code

```
adv_diff_cells_book.m
```

we consider a simple numerical experiment in which we compare two values of diffusivity. The two figures we present are constructed the same way. The top left panel shows the streamfunction and the initial distribution of tracer (the little black box). Subsequent panels show the tracer field at 0.144 dimensionless time unit increments. The tracer is capped at $C = 0.1$ because the diffusion leads to a reduction of the maximum value of tracer. The division between the two cells is indicated by a white, vertical line. In the absence of diffusion there would be no way for a tracer to move across the boundaries between cells.

Fig. 3.9 shows a case with a low diffusivity of $\nu = 10^{-3}$. The tracer is transported around the red (counter clockwise) cell roughly as one might expect. The role of diffusivity is elucidated if we compare Fig. 3.9 to Fig. 3.10. In the latter figure by the first time shown (the middle panel in the first row) some tracer appears in the top of the domain. This cannot be due to advection, and is due to diffusion and the periodic boundary conditions we choose to employ. The diffused tracer gets caught up in the upper branch of the cell and is advected leftward and downward, before finally being diffused out by the middle panel of the second row.

However, the moral of the story is plain to see: pure transport may move concentration but it cannot push material past the natural boundaries that occur in a flow. In the natural world, this means, for example, that a species of plankton in a cold core eddy may be transported hundreds of kilometers from its usual species range because the eddy remains a coherent fluid dynamical object over long time periods. The ever present turbulence of the natural world always provides some small scale diffusion, implying the eddy will not remain coherent forever. However, this natural diffusion is often dwarfed by that which adopted in numerical models. It is this "excess" diffusion that can move tracer across what would be natural transport barriers, as in

46 CHAPTER 3 Modeling active tracers

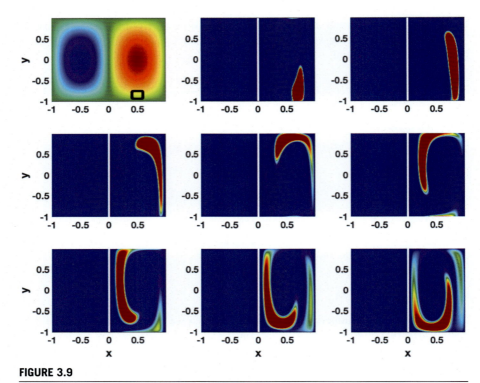

FIGURE 3.9

Top left panel shows the streamfunction and the initial distribution of tracer (black box). Subsequent panels show the tracer field at 0.144 dimensionless time unit increments. The tracer is capped at $C = 0.1$ and the division between the two cells is indicated by a white, vertical line. $\nu = 0.001$.

Fig. 3.10. This fact is vital to keep in mind in modeling exercises, and may suggest to the modeler that a better resolved "toy" model should be run along with a larger, more complete model that may have too much numerical diffusion.

3.7 Advection diffusion population model in 2D

We complete this chapter with a relatively simple model of an active tracer in two dimensions. We will consider the tracer from the previous section as a population that grows according to a logistic equation with a carrying capacity of 1. The equation governing the population, using the common short cut of denoting partial derivatives by subscripts, is thus

$$P_t = -uP_x - vP_y + \nu\nabla^2 P + P(1-P) \qquad (3.20)$$

3.7 Advection diffusion population model in 2D   47

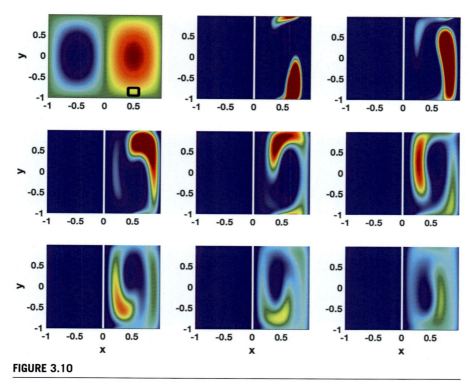

FIGURE 3.10

Top left panel shows the streamfunction and the initial distribution of tracer (black box). Subsequent panels show the tracer field at 0.144 dimensionless time unit increments. The tracer is capped at $C = 0.1$ and the division between the two cells is indicated by a white, vertical line. $\nu = 0.01$.

and the reader is cautioned that the parameter choices are for illustration and are not meant to be realistic in any way.

We repeat the experiment from the previous subsection. The logistic growth model is a switch in the

```
adv_diff_cells_book.m
```

code.

Comparing the population in Fig. 3.11 to the tracer in Fig. 3.9 with a low value of diffusivity, we conclude that the distribution of both fields is qualitatively similar. Examining the bottom right panel, we can note that the population field does have a larger areal extent of values larger than 0.1 (the red regions).

It is by comparing the population in Fig. 3.12 to the tracer in Fig. 3.10 with a high value of diffusivity that we can identify qualitative differences. Focusing on the leftmost panel in the second row we can see that for the population case, the small

48 CHAPTER 3 Modeling active tracers

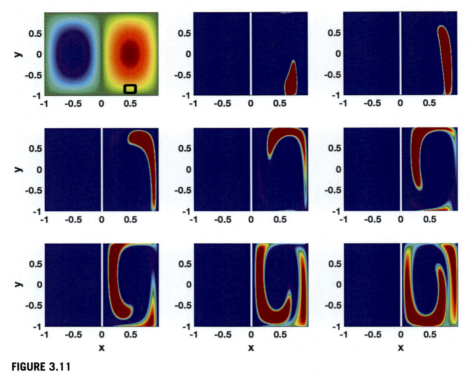

FIGURE 3.11

Top left panel shows the streamfunction and the initial distribution of tracer (black box). Subsequent panels show the population field at 0.144 dimensionless time unit increments. The population is capped at $C = 0.1$ and the division between the two cells is indicated by a white, vertical line. $\nu = 0.001$.

amount of material that was diffused across the bottom cell boundary to the top of the domain has not only been advected, but has also increased according to the logistic growth model. This growth has profound consequences since by the final time, shown in right panel in the bottom row, it has led to a situation where the vast majority of the domain has a population that is greater than 0.1. The perceptive reader will in fact note that it has in fact also led to diffusion across the right boundary, so that some population is evident near the left domain boundary.

The explorations in this chapter point to the subtle interplay of mechanisms in spatially distributed models involving biology. The distinction between waves and instabilities is fairly clear cut in hydrodynamics, and yet even in this context open problems remain. In coupled biophysical models, the number of mechanism grows to include diffusion and the reaction model, and hence the construction of appropriate toy problems to illustrate the interplay of mechanisms is of paramount importance. We will return to this issue after we introduce the reader to hydrodynamics.

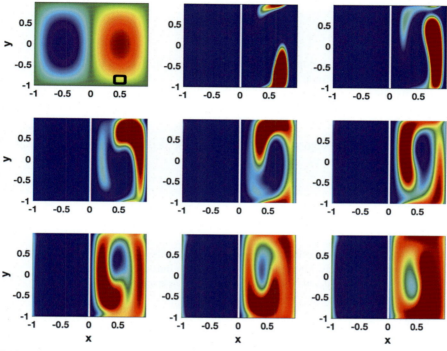

FIGURE 3.12

Top left panel shows the streamfunction and the initial distribution of tracer (black box). Subsequent panels show the population field at 0.144 dimensionless time unit increments. The population is capped at $C = 0.1$ and the division between the two cells is indicated by a white, vertical line. $\nu = 0.01$.

3.8 Mini-projects

1. How would you modify the spatially dependent Lotka-Volterra a model to have spatially dependent birth, death and predation rates? What would be the challenge numerically? Writing code would make this an extra challenging mini-project.
2. Modify the

    `adv_diff_lotka_volterra.m`

    code so that the population is governed by the time dependent Lotka-Volterra model from the previous chapter and discuss how the behavior changes.
3. Modify the

    `adv_diff_tandb.m`

    code so that phytoplankton is continuously released in a Gaussian shaped narrow region near the center of the domain.

50 **CHAPTER 3** Modeling active tracers

4. Modify the

```
adv_diff_cells.m
```

code so that the initial population is distributed as a line across the cell and discuss how the line is deformed by the flow.

5. Modify the

```
adv_diff_cells.m
```

code so that the population is governed by a time dependent streamfunction $\psi = cos(\omega t)\sin(\pi x)\cos(\pi y/2)$. Vary ω.

References

[1] P.K. Kundu, I.M. Cohen, Fluid Mechanics, 4th ed., Elsevier Academic Press, 2008.

[2] J.D. Logan, Nonlinear Partial Differential Equations, Wiley, 1994.

[3] James Dickson Murray, Mathematical Biology: I. An Introduction, Springer, 2002.

[4] G.B. Whitham, Linear and Nonlinear Waves, Wiley-Interscience, 1999.

CHAPTER

Fluid mechanics

4

CONTENTS

4.1 The basics of fluid mechanics: accounting for flow 51
4.2 Forces in a fluid: generalizing Newton's second law 53
4.3 The shallow water equations 57
4.4 Vorticity and stream function 62
4.5 Mini-projects 64
References 64

4.1 The basics of fluid mechanics: accounting for flow

The recognition that the motion of fluids is both aesthetically beautiful and practically useful has a long history in many cultures. The quantitative description of fluid motion, while more recent, has its Western roots in the Renaissance and is thus quite old in itself. In the language of the modern science we could say that fluid mechanics is a branch of classical physics, and hence dates back to Newton (who codified the fundamental laws) and Leibniz (who popularized differential calculus). Despite of this classification, it is a paradox of the modern division of the sciences that many students of physics never see a proper course in fluid mechanics. Much more often students of engineering encounter fluid mechanics in their undergraduate careers, while some students in the natural sciences see fluid mechanics in their graduate career (e.g. atmospheric science students and oceanographers).

The basic variable of fluid mechanics is the fluid velocity, which is a vector that depends on position and time; often written as $\vec{u}(\vec{x}, t)$. The basic mathematical conceit of fluid mechanics is the so-called Continuum Hypothesis which says that we may ignore the atomic nature of matter when describing fluids, and use the rules of ordinary differential and integral calculus. The Continuum Hypothesis is an example of what the theoretical physicist Eugene Wigner referred to as the "unreasonable effectiveness of mathematics in the natural sciences". The Continuum Hypothesis can be applied to a droplet that gathers on a rose bud after the rain, and to the motions of a spiral galaxy, a range of at length scales from 10^{-3} to 10^{20} m, which really does boggle the mind. The mathematician Leonard Euler came up with a useful version of the continuum hypothesis for the description of fluids. He defined a fluid particle as a small volume of fluid, that is large enough to have macroscopic properties (like temperature), but small enough so it may be mathematically treated as a singleton.

Physics and Ecology in Fluids. https://doi.org/10.1016/B978-0-32-391244-0.00014-0
Copyright © 2023 Elsevier Inc. All rights reserved.

52 CHAPTER 4 Fluid mechanics

This allows us to write a mathematical statement for the motion of a fluid particle,

$$\frac{d\vec{x}}{dt} = \vec{u}(\vec{x}(t), t) \tag{4.1}$$

which along with some initial conditions, say $\vec{x}(0) = \vec{a}$, would define the path a fluid particle takes. The solutions of this set of ordinary differential equations are called **pathlines**. On the face of it this is about as simple a mathematical statement as one can hope for. It is, however, not very useful as far as getting solutions as a "formula", or in what mathematicians call "closed form". This is because we do not know $\vec{u}$ and even if we did, we are evaluating it along the path (at $\vec{x}(t)$ for each time t). Mathematicians would say that this equation is strongly nonlinear, and would either approximate its solution (say using numerical methods) or try to prove general properties without finding actual solutions.

If we think about how we would go about measuring the properties of a fluid flow in the lab, the fluid particle may seem like an unnecessary complication. In practice, we would likely install a sensor at a particular location, and measure there. Measurements taken at fixed locations are called Eulerian measurements, after Leonard Euler. In contrast, measurements taken following a fluid particle are called Lagrangian measurement, after Joseph-Louis Lagrange.

Fig. 4.1 shows a practical example of the two types of measurement, which can give very different representations of the same physical phenomenon. For a theoretician looking to describe the motions of water in a lake, the Lagrangian point of view is impractical. This is because all of the tools for deriving the equations of fluid motion are designed for the Eulerian point of view (sometimes also called the "lab frame"). However any mathematical description of fluid flow must somehow account for the fact that fluids flow, and we can see how that affects the basic mathematical description of the rate of change in time.

Consider the temperature of a particular fluid parcel or particle (in the sense given above), $T(t, x(t), y(t), z(t))$. Since the location of the particle depends of time if we want to calculate the rate of change with respect to time, we must apply the Chain Rule from calculus:

$$\frac{d}{dt}\left[T(\vec{x}(t), t)\right] = \frac{\partial T}{\partial t} + \frac{\partial T}{\partial x}\frac{dx}{dt} + \frac{\partial T}{\partial y}\frac{dy}{dt} + \frac{\partial T}{\partial z}\frac{dz}{dt}.$$

Since the velocity of the fluid parcel is defined as

$$\vec{u} = (u, v, w) = \left(\frac{dx}{dt}, \frac{dy}{dt}, \frac{dz}{dt}\right) \tag{4.2}$$

we can define the so-called **material derivative**, which gives the rate of change following a fluid particle, as

$$\frac{DT}{Dt} = \frac{\partial T}{\partial t} + \vec{u} \cdot \nabla T. \tag{4.3}$$

4.2 Forces in a fluid: generalizing Newton's second law

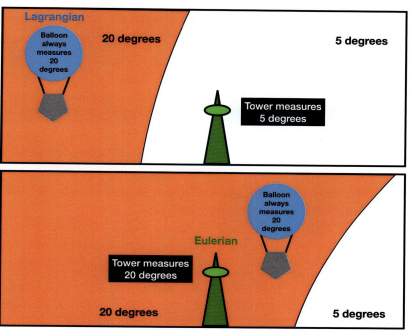

FIGURE 4.1

Cartoon showing the difference between Eulerian and Lagrangian measurements using a weather balloon and an observation tower as a warm front propagates past the tower.

As you can imagine, we will require this mathematical object in all descriptions that describe how a flowing fluid evolves in time. If you want a simple take away from this section, remember that to compute the rate of change of a fluid property you must account for changes in time and changes due to the fact that the fluid is flowing. Engineers call the latter ($\vec{u} \cdot \nabla T$ above) either the "advective" or "convective" terms (our personal preference is for "advective").

4.2 Forces in a fluid: generalizing Newton's second law

Newton's second law, which is the primary quantitative descriptor of particle motion in classical motion is often summarized as "$F = ma$". This is a bit misleading since it looks like a formula relating mass (m), acceleration (a) and force (F), when in fact it is a differential equation. To see that recall that for a particle the linear momentum is defined as

$$p = mv$$

54 **CHAPTER 4** Fluid mechanics

so that one way to express Newton's law is by saying that the rate of change of linear momentum is balanced by the sum of the forces, or mathematically

$$\frac{dm\vec{v}}{dt} = \sum_i \vec{F}_i. \tag{4.4}$$

The reader will notice that the mathematical objects in question are vectors. If we recall the definition of velocity as

$$\vec{v} = \frac{d\vec{x}}{dt}$$

and assume that mass does not change in time, we can write Newton's law in more familiar form as

$$m\frac{d^2\vec{x}}{dt^2} = \sum_i \vec{F}_i. \tag{4.5}$$

A concrete example would be a particle moving in the vertical, or z direction only under the action of gravity. The force of gravity will be written as $F_G = -mg$ where g is the acceleration due to gravity, so that Newton's law reduces to one ordinary differential equation (after we cancel the common factor of mass)

$$\frac{d^2\vec{x}}{dt^2} = -g. \tag{4.6}$$

To write a Newton's second law for fluids we have to ask "What are the relevant forces?" This leads to a larger consideration of forces unique to a continuum (usually covered in the field of continuum mechanics). Unfortunately this field is intertwined quite deeply in mathematical notation, and concepts that have a steep start up cost. However, it is possible to get the core ideas with only a minor loss of precision.

Consider two classical experiments. One is in fact a "non-experiment" since nothing happens. Consider a glass filled with water and ask what the force is on the bottom of the glass. If we neglect the air pressure above the surface of the water, we would say that the force per unit is called the fluid pressure, and must be due to the weight of the fluid above the bottom

$$F_H = -\rho g A h$$

where ρ is the fluid density (assumed constant for now), g is the acceleration due to gravity, A is the cross-section of the glass (assume it circular for concreteness) and h is the height of water in the glass. The subscript H is used to indicate that this is a hydrostatic state (i.e. no motion) and the minus comes from the convention that the z-axis points up, while gravity acts downward. If we consider the pressure at a point and ask ourselves how the pressure varies with z we can cancel the area A and write

$$p(h_2) - p(h_1) = -\rho g [h_2 - h_1],$$

4.2 Forces in a fluid: generalizing Newton's second law 55

so that dividing and taking the limit as $h_2 \to h_1$ gives the so-called hydrostatic pressure relation

$$\frac{dp_H}{dz} = -\rho g. \tag{4.7}$$

The second basic fluid experiment goes back to Newton, and consists of two large plates with a layer of fluid of thickness H in between. The bottom plate is held still, while the upper has a force acting on it so that the plate achieves a constant speed U in one direction (which we choose to coincide with the x axis). The fluid between the plates is measured to have a velocity that is only in the x direction and depends only on z, the vertical variable, so that

$$u = U\frac{z}{H},$$

which implies that $u = U$ at the upper plate. This means that each horizontal layer of fluid is dragged backwards by the layer below, so that starting with the speed U, the motion of the fluid is reduced to zero when z reaches 0. This type of force is called a shear force, and acts perpendicularly to the direction of motion.

In fact, for a fluid moving in a general way the forces inside the fluid, called continuum forces, or forces due to neighbors, or traction depend on where we are in the continuum, or $\vec{x}$ and what we consider as "outside" and "inside". Mathematically, if we label this force $\vec{t}$ we would write $\vec{t}(\vec{x}, \hat{n})$. The $\hat{n}$ is the mathematical terminology for the vector that points away from what we consider "inside" to what we consider "outside", or the normal vector. Writing things this way is sensible, but hard to work with mathematically. A famous theorem due to Augustin-Louis Cauchy allows us to write

$$\vec{t} = \mathbf{T}\hat{n}^T, \tag{4.8}$$

where $\mathbf{T}$ is called the stress tensor (or matrix) and $\hat{n}^T$ is the normal written as a column vector. This extraordinarily simple statement allows us to build an equivalent to Newton's second law using the material derivative from the previous section. Since the description is at a point, instead of mass we consider the density ρ, and for now we assume this quantity is constant and we write the fluid velocity as $\vec{u}(x, y, z, t)$ so that

$$\rho\frac{D\vec{u}}{Dt} = \vec{F}_G + \nabla \cdot \mathbf{T}. \tag{4.9}$$

The reason for the $\nabla\cdot$ term on the force due to neighbors is that while gravity is a body force, so acts over a whole volume, the force due to neighbors acts across an interface. The detailed derivation can be found in [1,2].

There is one thing people generally do with $\mathbf{T}$ and that is they write it to represent the two types of forces presented in the two basic experiments discussed above. This is done by expressing $\mathbf{T}$ as a sum of two terms. The term representing pressure acts equally in all directions, while the remaining piece represents all the possible forces

56 CHAPTER 4 Fluid mechanics

that aren't the same in all directions (this will contain the shear discussed in Newton's experiment above) like this

$$\mathbf{T} = -p\mathbf{I} + \sigma. \tag{4.10}$$

Here $\mathbf{I}$ is the identity matrix (1 along the diagonal and zero everywhere else). A tiny bit of algebra, where we choose the z axis to point "up" allows us to write

$$\rho\frac{D\vec{u}}{Dt} = -\nabla p + (0, 0, -\rho g) + \nabla \cdot \sigma, \tag{4.11}$$

and using the linear constitutive law of a Newtonian fluid we can simplify to the standard form of the Navier-Stokes equations

$$\rho\frac{D\vec{u}}{Dt} = -\nabla p + (0, 0, -\rho g) + \nu\nabla\vec{u} \tag{4.12}$$

where ν is a physical parameter referred to as the **kinematic viscosity** with the dimensions meters squared per second.

The Navier-Stokes equations are a very important set of equations, so much so that we will rewrite it in words:

The rate of change of linear momentum following a fluid particle is balanced by a sum of the pressure force, the gravity force, and all other (shear-type) forces due to neighbors.

The Navier-Stokes equations describe motions ranging form sub-centimeter to planetary scale motions. In practice, they are solved approximately using software, and software is limited by available memory. This means models that consider most environmental problems (think of a local reservoir) cannot resolve ALL scales in the problem. Often one imagines that there are large scale, coherent motions that we wish to concentrate on, and small scale, disordered (or turbulent) motions which we wish to pay less attention to. We will return to the mathematical details of this idea in Chapter 7. The above, heuristic description allows for an intuitive understanding of what most models actually do: they assume that the disordered (or turbulent) motions can be subsumed in a modified $\nabla \cdot \sigma$ term. In practice, this manifests itself as a diffusion term with an artificially high diffusion constant; the so-called **eddy viscosity**. Horizontal and vertical scales are often considered separately. To see why this makes sense, think of a reservoir. The horizontal length of this body of water may be a kilometer or two, while the depth may be ten meters or so, meaning that the ratio of vertical to horizontal length scale, or **aspect ratio** is much less than one. This implies that the artificial horizontal eddy viscosity should be much higher than the artificial vertical eddy viscosity. You can think of the eddy viscosity as representing what your model cannot tell you due to limited resolution.

If you prefer to think of the equation form of the above discussion, here are the Navier-Stokes equations as they might be applied to modeling scales larger than one meter of the reservoir introduced above,

$$\rho \frac{D\vec{u}}{Dt} = -\nabla p + (0, 0, -\rho g) + \mu_H \nabla_H^2 \vec{u} + \mu_V \partial_{zz} \vec{u}. \tag{4.13}$$

The two eddy viscosities have a subscript corresponding to the Horizontal and Vertical directions, so that $\mu_H \gg \mu_V \gg \mu_{molecular}$ and $\nabla_H^2 = \partial_{xx} + \partial_{yy}$. In the technical literature modeling of this type is often referred to as Reynolds Averaged or RANS. Very rarely is any discussion provided of when the eddy viscosity assumption may not be correct (even though this is ably discussed in standard text books, [2]), and the careful practitioner will consider models with different grid resolution to ensure results are independent of eddy viscosity values. As we will see next, a large branch of the historical study of lakes uses a set of equations that neglect viscosity altogether, or more accurately they only re-introduce eddy viscosity after the derivation is done, or *a posteriori*.

4.3 The shallow water equations

As mentioned above, the full set of Navier-Stokes equations, while theoretically complete, is rarely the set of equations used in practice. The need to get answers rapidly and with reasonable computational effort has motivated the development of a number of simplified models. In this section we develop an intuitive derivation of the so-called shallow water equations which form an essential part of the description of natural waters for which the aspect ratio is small.

We begin this section by summarizing our set of assumptions:

1. The density can be considered constant.
2. The horizontal length scale L is much larger than the vertical length scale of motion H, or $H \ll L$.
3. The pressure is hydrostatic (only due to the weight of the fluid above a point). This assumption also implies that the vertical component of acceleration can be disregarded (or put another way, the fluid moves in vertical columns).
4. The motion we are interested in can be described by the displacement of the lake's surface $\eta(x, y, t)$ and the vertically averaged velocity, i.e. we are not interested in how velocities vary in the vertical.
5. We do not need to consider the shear stresses so that all the forces in the fluid can be derived from the pressure p.

The basic situation is schematized in Fig. 4.2. The shallow water equations are discussed in general fluid mechanics books such as [2] as well as specialized geophysical fluid dynamics books such as [3]. While not necessary, in our exposition we consider the case with a flat bottom. This is done purely to make the algebra as simple as possible, since in our experience battling with algebra tires out readers, leaving no energy for critical examination of what portion of real-world physics is captured by the model and what portion is missing.

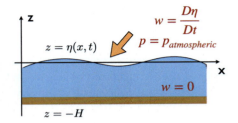

FIGURE 4.2

Summary of boundary conditions at the free surface.

As mentioned above, we will again adopt the philosophy of explaining the aspects of the mathematics that is relevant to the scientist, and leaving the details to more mathematically complete sources. Toward this goal, consider the question of what conditions must be satisfied at the surface. There are two conditions. One is basically common sense: **particles that start on the surface stay there**. To express this in equation form consider write down the equation for the surface

$$z = \eta(x, y, t)$$

and now compute the rate of change following a fluid particle (recalling that the variable on the left hand side of the equation is just the spatial coordinate z)

$$\frac{Dz}{Dt} = \frac{D\eta}{Dt}.$$

To simplify note that most terms in the material derivative of z are zero so that

$$\frac{Dz}{Dt} = w\frac{\partial z}{\partial z} = w$$

so that

$$w = \frac{D\eta}{Dt} \text{ at } z = \eta. \quad (4.14)$$

This boundary condition is called the kinematic boundary condition. It allows us to integrate the conservation of mass ($\vec{\nabla} \cdot \vec{u} = 0$) from the bottom (which we take to be flat and at $z = -H$) to the surface $z = \eta(x, t)$,

$$\int_{-H}^{\eta} u_x dz + \int_{-H}^{\eta} v_y dz + w(\eta) - w(-H) = 0.$$

Since there is no flow through the bottom $w(-H) = 0$ and from the kinematic or "particles on the surface stay there", boundary condition we find

$$\int_{-H}^{\eta} u_x dz + \int_{-H}^{\eta} v_y dz + \frac{D\eta}{Dt} = 0.$$

4.3 The shallow water equations 59

This can be further simplified by assuming that the horizontal velocity components are independent of depth (z). This allows us to take the u_x and v_y out of the integral and find

$$\frac{D\eta}{Dt} + (H+\eta)u_x + (H+\eta)v_y = 0.$$

Finally this can be simplified to give the conservation of mass

$$\frac{\partial \eta}{\partial t} + \nabla \cdot [(H+\eta)\vec{u}] = 0. \tag{4.15}$$

The first term represents the rate of change of the free surface with time while the second represents the horizontal flux of volume (and hence mass since the density is assumed constant). It states that in regions of horizontal velocity convergence the free surface rises, while in regions of divergence it falls. You can confirm this while washing dishes in your sink by clapping under the water. For both numerical methods and PDE theory the fact that (4.15) is in "conservation form" is important. As we will see this does not come about naturally for the momentum equations.

The second boundary condition at the surface is also common sense: the forces must not have a "jump" across the surface. To put this into mathematical form we need to recall that we assumed that only the pressure matters in specifying forces, so that the boundary condition becomes

$$p = p_{atmospheric} \text{ at } z = \eta. \tag{4.16}$$

This is called the dynamic boundary condition, and often the atmospheric pressure is neglected (basically because water is so much denser than air). To derive the momentum equations we begin with the vertical momentum equation. Since we assume motion is in vertical columns, we drop all terms except for the hydrostatic pressure terms, or

$$\frac{dp}{dz} = -\rho_0 g.$$

Integrating from an arbitrary height z^* to the surface $z = \eta$ gives

$$p(\eta) - p(x, y, z^*, t) = -\rho_0 g \eta(x, y, t) + \rho_0 g z^*. \tag{4.17}$$

But at the surface we know pressure has to match atmospheric pressure so that

$$p(z^*) = \rho_0 g \eta - \rho_0 g z^* - p_{atmospheric}. \tag{4.18}$$

If we let the horizontal gradient be given by

$$\nabla_H = (\partial_x, \partial_y)$$

we get

$$-\nabla_H p = -\rho_0 g \nabla_H \eta.$$

The horizontal momentum equations thus read

$$\rho_0 \frac{D(u,v)}{Dt} = -\nabla_H p$$

or using (4.18) and dividing by ρ_0

$$\frac{D(u,v)}{Dt} = -g\nabla_H \eta. \qquad (4.19)$$

This equation is not in "Conservation Form", though it is an exact representation for the conservation of horizontal momentum. It can be converted to be in conservation form using the Conservation of Mass and some clever manipulations.

There is no doubt that reducing the dynamics in a complex system like a lake or the coastal ocean to three equations is a nice result at the cost of only some mathematical trickery. It is worth revisiting however what was neglected in the process. Fig. 4.3 shows some of what is missing in diagram form. We have no means to put energy into our lake as the wind does, nor to take it out, as turbulent drag on the bottom would. Of course we can specify some initial state for a jet or other type of current, but if we want representations of wind and dissipative processes we must do this in an *ad hoc* manner.

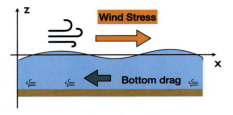

FIGURE 4.3

Real world processes that are neglected in the derivation of the shallow water equations that are often introduced on an *ad hoc* basis.

As a point of fact the "intuition" from the shallow water equations used for various aspects of lake physics often comes from an even simpler system, namely the linearized shallow water equations. When the linearization is about a background state of no current the momentum equations read

$$\frac{\partial \vec{u}}{\partial t} = -g\nabla \eta.$$

If we further assume that the bottom is flat, the linearized conservation of mass reads

$$\frac{\partial \eta}{\partial t} + H\nabla \cdot \vec{u} = 0.$$

4.3 The shallow water equations 61

We can now eliminate the velocity components by taking the divergence of the momentum equations and multiplying by H:

$$H\frac{\partial \nabla \cdot \vec{u}}{\partial t} = -gH\nabla^2\eta.$$

Next we take a time derivative of the conservation of mass and rearranging:

$$\frac{\partial^2\eta}{\partial t^2} = -H\nabla \cdot \frac{\partial \vec{u}}{\partial t}.$$

Eliminating the divergence of velocity term finally gives the classical wave equation (which is a hyperbolic equation)

$$\frac{\partial^2\eta}{\partial t^2} = c_0^2\nabla^2\eta \tag{4.20}$$

where the so-called shallow water speed is defined as

$$c_0 = \sqrt{gH}. \tag{4.21}$$

Shallow water waves are highly exceptional since they are non-dispersive (their wave speed does not depend on wavelength). Nevertheless, c_0 is widely used as a representative velocity (even though it is often far too high to represent true, naturally occurring velocities). The nondispersive nature of linear, shallow water waves also has important connotations for nonlinear shallow water theory, which are beyond our main narrative.

It is worth making a short note about why the linear theory worked out so well; namely that we were able to identify terms proportional to the divergence of the velocity field in both the conservation of mass, and a manipulation of the conservation of momentum equations. We will see that the situation is more complicated in the presence of rotation.

Finally, we can use the shallow water speed as a tool for classification of fluid flows. If we consider the ratio

$$Fr = \frac{U}{c_0}$$

where U is some estimate of typical flow velocities, we can see immediately that Fr is dimensionless. Let's assume that U is positive, and imagine that it represents the flow in a stream. If $Fr > 1$ waves cannot propagate to the left (or upstream). If $Fr < 1$ waves can propagate both upstream and downstream (though they would do so with speeds $c_0 - U$ and $c_0 + U$ due to the Doppler shift by the current.

Fr, the **Froude number** is one of the famous dimensionless parameters of fluid mechanics. Flows with $Fr < 1$ are called **subcritical** and flows with $Fr > 1$ are called **supercritical**. We will see Fr again when we discuss the role of rotation in the next chapter.

62   **CHAPTER 4** Fluid mechanics

4.4 **Vorticity and stream function**

We have worked quite hard to make material often presented in heavy mathematical form as intuitive as possible. Nevertheless there are a couple of mathematical ideas that merit a brief discussion. These both relate to the vector nature of the velocity field $\vec{u}$.

The first of these considers local rotation of fluid particles. This can be obtained using the curl operator of vector calculus, or in other words the cross-product of the gradient and the velocity field. We define the **vorticity** as

$$\vec{\omega} = \nabla \times \vec{u} \tag{4.22}$$

Readers who have taken a course in Vector Calculus can recall this quantity as a primary actor in the Stokes' theorem which tells us that, for a smooth orientable surface S bounded by a curve C,

$$\int\int_S \nabla \times \vec{u} \cdot d\vec{A} = \int_C \vec{u} \cdot d\vec{s}$$

or in other words that the net circulation around C is given by summing the contributions of vorticity over the surface.

Even if Stokes' theorem does not ring any bells, one can get intuition for vorticity using simplified configurations. So much of what we will discuss in this book concerns fluid motion that is nearly two dimensional, it is worthwhile working out the vorticity for motion in the $x - y$ plane for which $\vec{u} = (u(x, y, t), v(x, y, t), 0)$. A standard calculation shows that the vorticity has only a z component, i.e. $\vec{\omega} = (0, 0, \omega)$ and

$$\omega = \frac{\partial v}{\partial x} - \frac{\partial u}{\partial y}. \tag{4.23}$$

The above equation shows that even simple shear of the form $v = V(y)$ has non-negligible vorticity, and in particular if $v = \gamma y$ then $\omega = \gamma$. A positive vorticity corresponds to a local counterclockwise rotation.

In Fig. 4.4 we show how vorticity may be different even for two fluid particles moving along the same pathline. The vorticity thus provides information that the velocity field itself, and even the pathlines, do not. As we will see in Chapters 5 and 7 the vorticity is very useful when the effect of the Earth's rotation is accounted for.

The second useful mathematical quantity for two-dimensional flow is called the streamfunction, often labeled as ψ. It comes from the mathematical observation that if we have a two-dimensional flow for which the divergence vanishes ($\nabla \cdot \vec{u} = 0$) then we are guaranteed the existence of a function so that

4.4 Vorticity and stream function

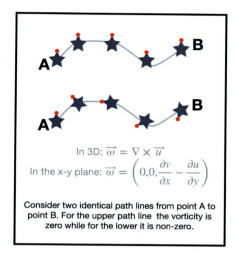

FIGURE 4.4

Cartoon illustrating how vorticity characterizes local rotation of fluid particles. Vorticity may differ even for particles moving along the same pathline.

$$(u, v) = \left(\frac{\partial \psi}{\partial y}, -\frac{\partial v}{\partial x} \right). \tag{4.24}$$

The flow is thus along the lines of constant ψ, as illustrated in Fig. 4.5.

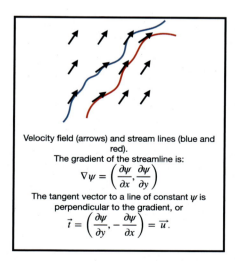

FIGURE 4.5

Cartoon illustrating how the streamfunction and velocity are related on the $x - y$ plane.

64   CHAPTER 4 Fluid mechanics

The streamfunction is a useful tool for some mathematical calculations, because it allows for the vorticity to be represented in terms of the streamfunction only, as opposed to the two components of velocity. Indeed, a quick calculation shows that

$$\omega = \frac{\partial v}{\partial x} - \frac{\partial u}{\partial y} = \frac{\partial^2 \psi}{\partial x^2} + \frac{\partial^2 \psi}{\partial y^2} = \nabla^2 \psi.$$

This fact will be used in Chapter 8.

4.5 Mini-projects

1. The classical theory of potential flow involves two-dimensional flows that are both incompressible ($\nabla \cdot \vec{u}$) and irrotational ($\nabla \times \vec{u}$). Using an appropriate reference (e.g. [2]) research the solution for potential flow past a sphere. Sketch the streamlines with and without circulation.
2. Using an appropriate reference (e.g. [2]) research vortex flows. Sketch the distribution of vorticity for an irrotational vortex, solid body rotation and a Rankine vortex.
3. Derive the equation governing the vorticity $\nabla \times \vec{u}$ for the shallow water equations model and discuss the physical meaning of all terms in the equation you have derived.
4. For small scale waves, like those excited by a water strider bug, surface tension modifies the boundary conditions at the fluid surface. Using an appropriate reference (e.g. [2]) discuss how the boundary conditions at the free surface are modified.
5. When the flow is steady, show that the streamlines and pathlines coincide (work in 2D to simplify the algebra). Is it true that if a flow is unsteady the pathlines and streamlines do not coincide?

References

[1] C.K. Batchelor, G.K. Batchelor, An Introduction to Fluid Dynamics, Cambridge University Press, 2000.
[2] P.K. Kundu, I.M. Cohen, Fluid Mechanics, 4th ed., Elsevier Academic Press, 2008.
[3] James C. McWilliams, Fundamentals of Geophysical Fluid Dynamics, Cambridge University Press, 2006.

CHAPTER 5

Rotating shallow water dynamics: An overview

CONTENTS

5.1	An overview of material	65
5.2	The effect of rotation: geostrophy	66
5.3	Rotating gravity waves in a channel: derivation	68
5.4	Rotating gravity waves in a channel: discussion	76
5.5	The circular lake: derivation	79
5.6	The circular lake: discussion	81
5.7	The tilted free surface problem	82
5.8	An instability problem: simulation	85
5.9	Mini-projects	89
	References	89

5.1 An overview of material

It is a famous truism in science that science does not seek to simply reproduce the complexity of nature, but it seeks to simplify in order to gain understanding. This chapter is all about a deceptively simple questions. One example is: "If we consider wind-induced motion in a lake on large scales (scales comparable to the basin itself) are there basic 'units' the motion can be broken up into?" We have seen that classical physics uses the concept of plane waves as a basic unit to understand wave motion, but the essential nature of side walls in a basin makes the basin problem considerably more challenging.

The first two thirds of this chapter build up a classical "model" in physical limnology: modes of free oscillation in a rotating basin. The model, while building on the plane wave idea, can be "machinery-heavy", meaning both in terms of notation and classical applied mathematics background. We thus provide a guide to key points. The reader familiar with some material can thus jump to sections containing material they feel is most important to them. The theoretical model is followed by a numerical simulation of the behavior of a tracer in a lake in which the dynamics start from a simple linear tilt. This can be thought of as the place where all the theory should make a natural appearance. It turns out that the natural world is more complex than the world based on linear mathematics and expressed in mathematical formulae. We provide examples of what really matters in the tilted surface case, namely the transport of tracer. In this way, we are taking the first steps to building a coupled biophysical, process study appropriate model.

66   CHAPTER 5   Rotating shallow water dynamics

The nonlinear simulations in this chapter have been carried out with the python codes:

```
sw2d_dg_circle_tracer.py
sw2d_dg_headlands_curved.py
```

corresponding to the simulations below in a circular domain and a periodic-in-x domain with a headland/peninsula protruding from the north side of the channel into the domain, respectively.

The chapter concludes with the second type of canonical problem in hydrodynamics, the problem of stability. Here, a basic, readily understood state is allowed to evolve and changes to a different, usually more complex, state. Again, we illustrate with the example of tracer transport.

1. The basic effect of rotation: geostrophy
2. Gravity waves in a channel: theoretically and practically
3. Basins: the classical solution in a circular lake, theoretically and practically
4. Basins: numerical solutions from a linear tilt of the free surface
5. The basics of instability
6. The instability of the double jet, numerically.

5.2 The effect of rotation: geostrophy

The shallow water equations discussed in the previous chapter are an excellent (though not perfect) tool for studying the large scale motions in the ocean and the atmosphere. This is because both the ocean and atmosphere form thin layers covering the large surface area of a (nearly) spherical Earth. The fundamental new bit of physics that needs to be incorporated into the shallow water equations as we wrote them is the fact that the Earth rotates, and thus our measurements on oceanic flows take place in a rotating reference frame. On the Earth we need to decide if we wish to deal with the whole spherical Earth or something more local so that we can choose Cartesian coordinates. In the spirit of the simplest possible, viable model, we will choose Cartesian coordinates and consider only the component of rotation aligned with the local normal. This is a fairly standard approximation of the spherical planet and goes under the name of the f plane ([4,5]). The resulting equations read

$$\frac{D\vec{u}}{Dt} + (-fv, fu) = -g\nabla_H \eta$$

$$\frac{\partial \eta}{\partial t} + \nabla \cdot [(H + \eta)\vec{u}] = 0. \tag{5.1}$$

Here $f = 2\Omega \sin\theta$ where Ω is the frequency of the Earth's rotation and θ is the latitude. For mid-latitudes the typical value of f is 1×10^{-4} $1/s$. Notice also that the formula for f implies that rotation effects lose importance as one approaches the equator.

5.2 The effect of rotation: geostrophy

Our first task in determining the simplest set of equations when rotation dominates, will be to scale the equations. Let $\vec{u} = U\tilde{u}$, $\eta = He$, $(x, y) = L(\tilde{x}, \tilde{y})$ and choose the **advective time scale** $T = L/U$. Notice that we did not choose the same length scale for x and the free surface height η. This is sensible since x might vary on the scale of tens of kilometers while η may be a meter or so. The momentum equations thus read

$$\frac{U^2}{L}\frac{D\tilde{u}}{Dt} + fU(-\tilde{v}, \tilde{u}) = -g\frac{H}{L}\tilde{\nabla}e$$

and we rearrange so that they read

$$\frac{U}{fL}\frac{D\tilde{u}}{Dt} + (-\tilde{v}, \tilde{u}) = -g\frac{H}{LfU}\tilde{\nabla}e.$$

The dimensionless number

$$Ro = U/fL \tag{5.2}$$

is called the **Rossby number**, and measures the typical ratio between the inertia terms and the rotation terms. As we mentioned above, f is a small parameter, on the order of 10^{-4} s^{-1} for mid-latitudes, but the length scales associated with rotation are long (more than 10 km), and moreover the velocity scales are small, on the order of a few centimeters a second. Thus $Ro \ll 1$ is a reasonable assumption and the acceleration and inertia terms are negligible. In order to have any sort of non-trivial balance we need the pressure gradient terms on the right hand side to not vanish and this gives us a way to choose H, the scale of η,

$$\frac{gH}{fLU} = 1.$$

Thus the approximate equations read

$$(-\tilde{v}, \tilde{u}) = -\tilde{\nabla}e$$

or, as they are often presented, re-dimensionalized to read

$$f(-v, u) = -g\nabla\eta. \tag{5.3}$$

These equations are called **geostrophic balance**. If you take the dot product of (5.3) with $\vec{u}$ you see that $0 = -g\vec{u} \cdot \nabla\eta$, or in other words that velocity is along lines of constant η. This is amazingly different from how fluids behave on our scales of experience, where flow is generally from high to low pressure. It is for this reason that large scale weather maps on television and the web show you lines of constant pressure, since as a first approximation the flow is along (and not across) these lines.

One aspect of the above calculation may have troubled you; namely the choice of scale for η in the assumption

$$\frac{gH}{fLU} = 1.$$

CHAPTER 5 Rotating shallow water dynamics

What if we didn't do this? Well, we would need to define another dimensionless number. The traditional way to do this is to notice that the numerator in the above expression is the square of the non-rotating shallow water wave speed (remember shallow water waves without rotation are non-dispersive) so if we say $c_0 = \sqrt{gH}$ then the left hand side of the expression reads

$$\frac{c_0^2}{fLU}$$

which we can multiply on the top and bottom by U and get

$$\frac{c_0^2}{U^2}\frac{U}{fL}.$$

We can recognize the second combination of parameters as the Rossby number and note immediately that the first fraction is also dimensionless. Traditionally the ratio of typical velocities to the typical wave speed is called the Froude number, written as

$$Fr = \frac{U}{c_0}$$

so that for low speed flows $Fr \ll 1$. In any case this definition allows us to write the dimensionless equations in a more general way as

$$Ro\frac{D\vec{u}}{Dt} + (-v, u) = -\frac{Ro}{Fr^2}\nabla e \tag{5.4}$$

where I have dropped the tildes.

5.3 Rotating gravity waves in a channel: derivation

It is important to understand the impact of rotation (i.e. the Coriolis force on the f-plane) on shallow water waves to build intuition on the behavior of waves in lakes. An example of a classical, analytically tractable problem is that of free linear wave oscillations in a periodic, or infinite, channel. This problem involves boundaries, and thus poses some mathematical difficulties. To set the stage, let's consider the problem of free waves.

In the last chapter we showed that in the absence of rotation the linearized shallow water equations are equivalent to a classical wave equation. They are thus hyperbolic waves, and non-dispersive. In the presence of rotation this is no longer case. The linearized momentum equations read

$$\frac{\partial \vec{u}}{\partial t} + (-fv, fu) = -g\nabla\eta.$$

5.3 Rotating gravity waves in a channel: derivation

We again assume that the bottom is flat, so that the linearized conservation of mass matches that in the non-rotating case, namely

$$\frac{\partial \eta}{\partial t} + H\nabla \cdot \vec{u} = 0.$$

The reduction to a single equation is non-trivial because of the presence of the $(-fv, fu)$ term. There are three possible ways around the problem

1. Derive a single equation via some tricky, and not very general, algebra.
2. Use the most general decomposition of the vector field $\vec{u}$ and derive equations for each part.
3. Use the wave ansatz to solve for plane waves solutions of the whole system.

The third seems like the most reasonable compromise between ease of presentation and generalizability.

We rewrite the system of governing equations in a slightly different form, with all the time derivatives on one side and all the space derivatives on the other

$$\begin{pmatrix} u \\ v \\ \eta \end{pmatrix}_t = \begin{pmatrix} fv - g\eta_x \\ -fu - g\eta_y \\ -Hu_x - Hv_y \end{pmatrix}.$$

Recall, that we have assumed that H is constant, meaning that all the parameters are constants as opposed to functions of space. We now make what is called the "wave ansatz", or the plane wave solution,

$$(u, v, \eta) = (u_0, v_0, a_0) \exp[i(kx + ly - \sigma t)].$$

If we let the vector of coefficients be called $\vec{w} = (u_0, v_0, a_0)$ then substitution converts the partial differential equations (PDEs) into a matrix problem $\sigma I \vec{w} = A\vec{w}$ or

$$(A - \sigma I)\vec{w} = 0,$$

where

$$A = \begin{pmatrix} 0 & if & kg \\ -if & 0 & lg \\ kH & lH & 0 \end{pmatrix}. \tag{5.5}$$

We would like to guarantee a non-trivial solution for the amplitudes $\vec{w}$. How do we do that? Well if we specify the wavenumber (k, l) and the physical parameters (H, f, g) then a non-trivial solution is guaranteed provided

$$\det(A - \sigma I) = 0$$

which is the same condition as the one that allows us to find the eigenvalues of A. Thus the dispersion relation is given by the eigenvalues of the matrix A and

CHAPTER 5   Rotating shallow water dynamics

the amplitudes by the eigenvectors. Taking the determinant gives the **characteristic polynomial** (it is a good idea to do this yourself)

$$-\sigma^3 + \sigma g H(k^2 + l^2) + \sigma f^2 = 0$$

which gives the trivial root $\sigma = 0$ and the two waves ("rightward" and "leftward" propagating)

$$\sigma = \pm\sqrt{f^2 + gH(k^2 + l^2)}.$$

If we recall from the non-rotating case that the non-dispersive wave speed as $c_0 = \sqrt{gH}$ we can rewrite the dispersion relation as

$$\sigma(k, l) = \pm\sqrt{f^2 + c_0^2(k^2 + l^2)}. \tag{5.6}$$

Waves with this dispersion relation are called rotation modified gravity waves, or Poincaré waves. What about the trivial root? Well saying that $\sigma = 0$ is equivalent to saying that the time derivative terms are identically zero, and this is exactly what our scaling analysis showed for the slow, or geostrophic motion.

Of course if wanted the amplitudes of the individual components themselves we would now find all the eigenvectors. We leave this as one of the mini-projects for this chapter.

Are Poincaré waves dispersive? Consider the simplest case where the waves propagate to the right along the x axis, so that

$$\sigma = \sqrt{f^2 + c_0^2 k^2}.$$

The frequency clearly depends on the wavelength in a non-trivial way. Poincaré waves are thus dispersive waves (and not governed by a hyperbolic equation like the classical wave equation). We can delve into this a bit more deeply by computing the two wave speeds.

The phase speed (the speed of individual crests) is defined as

$$c_p = \frac{\sigma}{k}$$

and using the dispersion relation we find

$$c_p = \sqrt{f^2/k^2 + c_0^2}.$$

It is interesting that the phase speed can grow without bound as k tends to zero (i.e. the limit of long waves). This seems a contradiction of the physical fact that the speed of light sets the upper bound on any possible wave propagation. To resolve this conundrum, recall that the group speed is defined as

$$c_g = \frac{d\sigma}{dk}$$

5.3 Rotating gravity waves in a channel: derivation

and tells us the speed with which energy propagates. Using the dispersion relation we find

$$c_g = \frac{c_0^2 k}{\sqrt{f^2 + c_0^2 k^2}}.$$

As k tends to zero (i.e. the limit of long waves) a simple calculation shows that c_g tends to zero as well. The limit of short waves is a bit less interesting because the derivation of shallow water theory assumes that waves are longer than the typical depth, nevertheless the math can be done. We find that as $k \to +\infty$ both c_p and c_g tend to c_0. This makes sense, since it says that rotation effects do not matter very much for short waves.

So far we have been quite careful to avoid any side boundaries. Before we look at boundaries in more generality, we present an old solution due to Lord Kelvin of practical importance. Imagine we have a coastline at $x = 0$ running North-South. Kelvin had the idea to look for wave solutions that propagate North-South, but have a "to-be-determined" structure in the East-West direction. For concreteness think of waves in the Pacific near the west coast of North America.

He started his calculations with the inspired guess of setting $u = 0$. This meant that he never had to deal with any boundary conditions! Sometimes, the great ideas are so simple. The governing equations with this assumption, and using subscripts for partial derivatives for convenience, read

$$-fv = -g\eta_x$$

for the East-West momentum,

$$v_t = -g\eta_y$$

for the North-South momentum, and

$$\eta_t + Hv_y = 0$$

for the mass. We can take the time derivative of the East-West momentum equation, multiply the North-South momentum equation by f and add the two equations to get

$$0 = -g\eta_{xt} - gf\eta_y.$$

Divide by g and assume $\eta = \exp[i(ly - \sigma t)]\phi(x)$ to get

$$0 = i\sigma\phi'(x) - fil\phi(x)$$

or

$$\phi'(x) = \frac{fl}{\sigma}\phi(x)$$

which has the solution

$$\phi(x) = A\exp(flx/\sigma),$$

72 CHAPTER 5 Rotating shallow water dynamics

where $A = 1$ without loss of generality since the theory is linear. We thus have

$$\eta = \cos(ly - \sigma t) \exp(flx/\sigma).$$

To find $\sigma(l)$ we take the y derivative of the North-South momentum equation and multiply by $-H$ to find

$$-Hv_{ty} = gH\eta_{yy}.$$

We then take a t derivative of the conservation of mass equation to find

$$\eta_{tt} = -Hv_{ty}$$

and eliminating v_{ty} gives the wave equation

$$\eta_{tt} = gH\eta_{yy}.$$

But if we define $c_0 = \sqrt{gH}$ this means

$$\sigma(l) = c_0 l.$$

We can thus write the solution for the free surface as

$$\eta = \cos[l(y - c_0 t)] \exp(flx/\sigma)$$

which we can rewrite in a somewhat more clever way. Recall that $\sigma/l = c_0$ so that we can define a decay scale

$$L_{decay} = \frac{c_0}{f}$$

and write then solution as

$$\eta = \cos[l(y - c_0 t)] \exp(x/L_{decay}) \tag{5.7}$$

valid for $x \leq 0$. How far from the wall the Kelvin wave has an appreciable amplitude is thus dependent on f, which changes with latitude but is typically around 1×10^{-4} s^{-1} for mid-latitudes, and is never larger than 1.15 times this value.

What is the physical content of this solution? Well it tells us that in the East-West direction the largest wave amplitude is right at the coast, and the wave decays exponentially as we go westward. In the North-South direction the wave is actually non-dispersive and propagates with a phase and group speed equal to $c_0 = \sqrt{gH}$, the wave speed of non-rotating shallow water waves. This wave is called the Kelvin wave, after its discoverer, and is said to "lean rightward on the coast" in the northern hemisphere. As an exercise, you can think about Kelvin waves in the southern hemisphere ($f < 0$) or ones for which the coast is on the left (like on the east coast of North America).

With the above simple cases in hand, we are ready to look for analytical solutions to the linearized rotating shallow water equations in a north-south running channel.

5.3 Rotating gravity waves in a channel: derivation

Mathematically this channel is given by $[0, L_x] \times [-\infty, \infty]$. The governing equations are

$$u_t - fv = -g\eta_x \,, \tag{5.8}$$
$$v_t + fu = -g\eta_y \,, \tag{5.9}$$
$$\eta_t = -Hu_x - Hv_y \,. \tag{5.10}$$

We look for solutions in what is called "normal mode form", or as "normal modes". This means the form is

$$\begin{pmatrix} u \\ v \\ \eta \end{pmatrix} = e^{i(ly - \sigma t)} \begin{pmatrix} \hat{u}(x) \\ \hat{v}(x) \\ \hat{\eta}(x) \end{pmatrix} \,, \tag{5.11}$$

where σ is a real number found as part of the solution, and $l = 2\pi m / L_y, m = 1, 2, \ldots$ since the channel is periodic and bounded in y. It is worth noting that this is similar, though slightly more complicated than what we did in the case of plane waves. Under the ansatz (5.11), the system (5.8)–(5.10) reduces to an eigenvalue problem again, but now this eigenvalue problem is a differential equation

$$\begin{pmatrix} 0 & f & -g\frac{d}{dx} \\ -f & 0 & -gil \\ -H\frac{d}{dx} & -Hil & 0 \end{pmatrix} \begin{pmatrix} \hat{u} \\ \hat{v} \\ \hat{\eta} \end{pmatrix} = -i\sigma \begin{pmatrix} \hat{u} \\ \hat{v} \\ \hat{\eta} \end{pmatrix} \,. \tag{5.12}$$

The boundary conditions are given by no normal flow through the channel walls, i.e.,

$$\hat{u} = 0 \quad \text{at} \quad x = 0, L_x \,. \tag{5.13}$$

Suitable conditions on $\hat{v}$ and $\hat{\eta}$ can be determined by substituting $\hat{u}(x = 0, L_x) = 0$ into the eigenvalue problem. After some light algebra, we recover the Robin type boundary conditions

$$fl\hat{\eta} - \sigma\hat{\eta}_x = 0 \quad \text{at} \quad x = 0, L_x \,, \tag{5.14}$$
$$fl\hat{v} - \sigma\hat{v}_x = 0 \quad \text{at} \quad x = 0, L_x \,. \tag{5.15}$$

Eliminating $\hat{u}$ and $\hat{v}$ from the eigenproblem (5.12) gives the single equation for $\hat{\eta}$

$$\hat{\eta}_{xx} + \left(\frac{\sigma^2 - f^2}{c_0^2} - l^2 \right) \hat{\eta} = 0 \,, \tag{5.16}$$

where $c_0 = \sqrt{gH}$. What we have found has a mathematical name (and hence a well-developed mathematical theory: the Sturm-Liouville differential eigen-problem, which may be solved by considering separate cases based on the sign of the quantity in parentheses in (5.16). Many classical calculations in environmental fluid dynamics follow the pattern of reducing problems to classical mathematical forms. Today, since

74   **CHAPTER 5**   Rotating shallow water dynamics

many students do not take classical mathematics courses, the theory ends up having the feel of something written in a lost or ancient language.

Regardless of how one feels about the reduction presented above it does allow us to proceed in a systematic, case by case manner.

Case I: $\frac{\sigma^2 - f^2}{c_0^2} - l^2 \geq 0$

In this case, the eigen-problem can be written as

$$\hat{\eta}_{xx} + \lambda^2 \hat{\eta} = 0\,, \tag{5.17}$$

where

$$\lambda = \sqrt{\frac{\sigma^2 - f^2}{c_0^2} - l^2}\,. \tag{5.18}$$

The general solution is

$$\hat{\eta} = A \sin(\lambda x) + B \cos(\lambda x)\,. \tag{5.19}$$

Imposing the boundary condition (5.14) at $x = 0$ yields

$$A = \frac{fl}{\sigma \lambda} B\,, \tag{5.20}$$

hence,

$$\hat{\eta} = B \left[\frac{fl}{\sigma \lambda} \sin(\lambda x) + \cos(\lambda x) \right]\,. \tag{5.21}$$

Imposing (5.14) at $x = L_x$ gives the equation

$$\left(f^2 l^2 + \sigma^2 \lambda^2 \right) \sin(\lambda L_x) = 0\,, \tag{5.22}$$

for which there are a multitude of possible roots. First, assume the quantity multiplying $\sin(\lambda L_x)$ is zero, i.e.,

$$f^2 l^2 + \sigma^2 \left(\frac{\sigma^2 - f^2}{c_0^2} - l^2 \right) = 0\,, \tag{5.23}$$

which may be rewritten as a quadratic equation in powers of σ^2,

$$\sigma^4 - (f^2 + c_0^2 l^2)\sigma^2 + f^2 l^2 c_0^2 = 0\,. \tag{5.24}$$

Thus, the roots are given by

$$\sigma^2 = \frac{f^2 + c_0^2 l^2 \pm \sqrt{\left(f^2 - c_0^2 l^2 \right)^2}}{2}\,. \tag{5.25}$$

5.3 Rotating gravity waves in a channel: derivation

Therefore,

$$\sigma^2 = f^2, \quad \sigma^2 = c_0^2 l^2.$$
(5.26)

However, these roots are unphysical for eigenfunctions of the form (5.19) since they break the assumption that λ is real and positive (in other words these waves would either shrink to zero or grow without bound). Hence they must be discarded.

Returning to Eq. (5.22), the other possible roots are given by the zeros of the sine functions, that occur for

$$\lambda L_x = n\pi, \quad n = 0, 1, \cdots.$$
(5.27)

Squaring this relation and substituting the expression for λ gives

$$\sigma^2 = f^2 + c_0^2(k^2 + l^2),$$
(5.28)

where

$$k = \frac{n\pi}{L_x}, \quad n = 0, 1, \cdots,$$
(5.29)

represent the across-channel wavenumber. The dispersion relation (5.28) is the classical result for Poincaré (rotating gravity) waves with only certain values of k possible. It is worth mentioning that the case where $\lambda = 0$ is recovered when $k = 0$, a wave with constant cross-channel structure.

Case II: $\frac{\sigma^2 - f^2}{c_0^2} - l^2 < 0$

In this case, the eigenproblem can be written as

$$\hat{\eta}_{xx} - \lambda^2 \hat{\eta} = 0,$$
(5.30)

where

$$\lambda = \sqrt{l^2 - \frac{\sigma^2 - f^2}{c_0^2}}.$$
(5.31)

The general solution is

$$\hat{\eta} = C \sinh(\lambda x) + D \cosh(\lambda x).$$
(5.32)

Imposing the boundary condition (5.14) at $x = 0$ yields

$$C = \frac{fl}{\sigma \lambda} D,$$
(5.33)

as before. Hence,

$$\hat{\eta} = D \left[\frac{fl}{\sigma \lambda} \sinh(\lambda x) + \cosh(\lambda x) \right].$$
(5.34)

76   **CHAPTER 5**   Rotating shallow water dynamics

Imposing (5.14) at $x = L_x$ gives

$$\left(f^2 l^2 - \lambda^2 \sigma^2\right) \sinh(\lambda L_x) = 0. \tag{5.35}$$

Since $\sinh(\lambda L_x)$ cannot be zero $(L_x, \lambda > 0)$, it follows that the eigenvalues are given by the equation

$$f^2 l^2 + \sigma^2 \left(\frac{\sigma^2 - f^2}{c_0^2} - l^2\right) = 0, \tag{5.36}$$

which was solved in Case I,

$$\sigma^2 = f^2, \quad \sigma^2 = c_0^2 l^2. \tag{5.37}$$

In this case, these values of sigma do not yield a contradiction, and the following eigenfunctions are recovered in each case

$$\sigma = f \Rightarrow \lambda = \frac{f}{c_0} \Rightarrow \hat{\eta} = E e^{-\lambda(L_x - x)}, \tag{5.38}$$

$$\sigma = -f \Rightarrow \lambda = \frac{f}{c_0} \Rightarrow \hat{\eta} = D e^{-\lambda x}, \tag{5.39}$$

$$\sigma = c_0 l \Rightarrow \lambda = \frac{f}{c_0} \Rightarrow \hat{\eta} = E e^{-\lambda(L_x - x)}, \tag{5.40}$$

$$\sigma = -c_0 l \Rightarrow \lambda = \frac{f}{c_0} \Rightarrow \hat{\eta} = D e^{-\lambda x}. \tag{5.41}$$

Here, we introduced $E = D e^{\lambda L_x}$ for ease of interpretation. The quantity c_0/f is referred to as the *external Rossby deformation radius*, and represents the decay length scale of these coastally trapped eigenmodes. Here, the $\sigma = \pm f$ solutions correspond to the so-called *inertial oscillation* and the $\sigma = \pm c_0 l$ solutions correspond to the *Kelvin wave*. Kelvin waves possess the property that their frequency can be less than the inertial frequency, and for closed basins this feature is typically used to identify which modes are Kelvin modes ([9]).

5.4 Rotating gravity waves in a channel: discussion

It is worth taking a moment to survey what we have learned. We certainly now know something about the possible variety of wave motions in the channel; almost as if we were cataloging the life forms found in the channel. However, the theory does not really tell us what we should expect to see occurring in the channel for a particular set of initial conditions; in our life form analogy we have not learned what the life-forms actually do. The theory does put some restriction on the qualitative features, e.g. propagating features whose amplitude is largest near the channel walls should correspond to Kelvin waves. Part of this is due to the assumption of linearity, which

5.4 Rotating gravity waves in a channel: discussion

makes all amplitudes equal in the eyes of the theory. In the physical world a light breeze over the lake and a massive storm system should lead to a very different response. The sharp eyed reader will note the theory is even more restrictive than that: the mathematical procedure that yields simplification restricts which variable is being examined in detail and hence only the dispersion relation giving frequency as a function of wave number and the modal structure for the free surface are actually given by the theory. The remaining variables of interest, e.g. the velocity components, are recovered *a posteriori*. Thus theory is a fairly heavy ask for the practically minded scientist, an issue that we will revisit once we have both some more theory under our belt and a numerical simulation to compare against.

In the interest of fairness, we can return to something we said earlier, and use the theory to learn something about motions inside lakes. Most temperate lakes are stratified for a significant portion of the year, and for deep lakes this stratification often consists of a thin upper layer and a thicker lower layer. The theory we have derived can be applied to internal motions under the so-called reduced gravity assumption: we assume that the internal interface is the only active interface and has a density jump $\Delta \rho$ across it. The reduced acceleration due to gravity (often just called the reduced gravity) reads $g' = \Delta \rho g$ and the entire theory can be recycled except for the change from g to g'. But what a big change it is! This is because the wave speed now reads

$$c_{internal} = \sqrt{g'H} \ll c_0$$

and because the density change (scaled by a reference density around $1000 \, \mathrm{kg \, m^{-3}}$) is typically less than 0.01 we have that this speed for internal motions is much smaller than for the motions of the free surface (sometimes these are called baroclinic and barotropic speeds, respectively).

Now we follow the chain of implications from this relatively simple fact. We recall that for Kelvin waves the decay scale reads

$$L_{decay} = \frac{c_0}{f}$$

which for internal Kelvin waves would have to be changed to

$$L_{decay} = \frac{c_{internal}}{f}.$$

Thus the decay scale of internal Kelvin waves is much smaller. Moreover, it means that surface Kelvin waves are only features of very large lakes, while internal Kelvin waves can be measured in lakes of modest size ([1,7]).

We demonstrate this effect in Figs. 5.1 and 5.2, where the latter shows the result for the internal waves. Here we have scaled the x variable by the channel width. It can be seen that the surface Kelvin wave decays very little across the channel, while the internal Kelvin wave has a nearly zero amplitude at the channel mid-depth. The two figures also show two Poincaré wave modes (the first and fourth). The differences

CHAPTER 5 Rotating shallow water dynamics

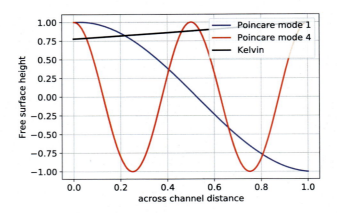

FIGURE 5.1

Sample profiles for two Poincaré waves and the Kelvin wave for surface, or barotropic waves in a channel.

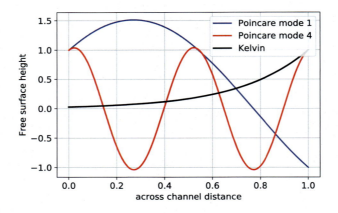

FIGURE 5.2

Sample profiles for two Poincaré waves and the Kelvin wave for internal, or baroclinic, reduced gravity waves in a channel.

between the surface and internal cases are not as pronounced for these, though the frequencies for the two cases would be very different.

Though indirect this result is useful, because it tells the field scientist what waves NOT to expect for mid-sized lakes (surface Kelvin waves). Nevertheless, the finish line takes considerable algebra-based work to get to, and many journal articles do not have the space to explain the details of the mathematical calculation or assumptions before applying the result.

Indeed, a modern, computational scientist has options. They can use a good software model tailored to idealized situations to actually solve initial value problems, or perhaps a more complete lake model to look at a realistic representation of the motions we have described theoretically above.

However, it is worth noting a key challenge. Since all realistic models will invoke some form of an ad hoc eddy viscosity, the simulated waves cannot ever propagate without changing amplitude, the way the theory in the previous section predicted. In other words, realism comes at a price, and it is important to gauge how well a given model does prior to applying it to a real world problem. This is the real utility of idealized mathematical solutions, they give us test cases for which we know the answer and these can in turn give us confidence in numerical models.

5.5 **The circular lake: derivation**

The case of the circular lake represents the simplest closed basin, as well as a classical solution of applied mathematics dating back to the late 19th century. This passage of time, however, has not meant that the presentation of the solution has been improved substantially.

We consider polar coordinates, with

$$\vec{u} = u^{(r)}\hat{r} + u^{(\theta)}\hat{\theta}$$

where the hats denote basis vectors in the radial and azimuthal directions. Rewritten in these coordinates, and in terms of these velocity components, the governing equations read

$$u_t^{(r)} - fu^{(\theta)} = -g\eta_r \, , \tag{5.42}$$

$$u_t^{(\theta)} + fu^{(r)} = -g\frac{1}{r}\eta_\theta \, , \tag{5.43}$$

$$\eta_t = -Hu_r^{(r)} - H\frac{u^{(r)}}{r} - H\frac{1}{r}u_\theta^{(\theta)} \, . \tag{5.44}$$

If we assume separable solutions (written here for the free surface η)

$$\eta(r, \theta, t) = \exp(i\omega t)\cos(n\theta)G(r) \tag{5.45}$$

where n gives the mode number in the azimuthal (or around the lake) direction. Under this assumed separable structure, the no-normal flow boundary condition at the outer circular wall is transformed to

$$i\omega\frac{\partial\eta}{\partial r} + \frac{f}{r}\frac{\partial\eta}{\partial\theta} = 0 \, , \qquad \text{at} \qquad r = R \tag{5.46}$$

where R is the radius of the lake. This bizarre-looking boundary condition couples the eigenvalue ω (the frequency of the oscillation) to the azimuthal mode number, n. This

80 **CHAPTER 5** Rotating shallow water dynamics

coupling, unfortunately, terribly complicates the problem, mathematically speaking! In our opinion, more mathematical texts should be up front about what bits of tricky algebra actually lead to problems down the line!

Carrying on, the eigenvalue problem for the radial structure function $G(r)$ may be written as

$$\frac{d^2 G}{dr^2} + \frac{1}{r}\frac{dG}{dr} + \left(\frac{\omega^2 - f^2}{gH} - \frac{n^2}{r^2}\right)G = 0, \tag{5.47}$$

along with the corresponding outer boundary condition

$$\omega r \frac{dG_n}{dr} + nf G_n = 0, \qquad \text{at} \qquad r = R \tag{5.48}$$

Here n is an integer parameter that effectively counts the modes, and for mathematically savvy readers the ODE may be recognized as either the Bessel equation (when $|\omega| > |f|$) or the modified Bessel equation (when $|\omega| < |f|$) ([8]). Unfortunately even though the governing equation is a well-studied differential equation, the frequency ω cannot be solved for exactly. A variety of classification results have been derived instead. Waves for which $|\omega| > |f|$ are called either super-inertial or Poincaré waves, and waves for which $|\omega| < |f|$ are called either sub-inertial or Kelvin waves. Sub-inertial waves only exist in a lake with a large enough radius R (this was derived by Lamb in the early 20th century, but see the more modern references in [2,9]) when

$$S = \frac{c_0}{fR} < \sqrt{\frac{1}{n(n+1)}} \tag{5.49}$$

The above suggests sub-inertial or Kelvin waves are more likely to be features of large lakes. In contrast, super-inertial waves can be calculated in all cases; even a pond has a number of different modes of oscillation. For internal modes we have

$$S = \frac{c_{internal}}{fR} < \sqrt{\frac{1}{n(n+1)}} \tag{5.50}$$

and since the only change is a decrease of the numerator, like in the channel, internal Kelvin waves can occur in mid-sized lakes ([1,7]).

Examples of the free surface manifestation of Kelvin wave modes are presented in Fig. 5.3. It can be seen that the waves are trapped near the coasts and increasing n means more repetitions in the azimuthal direction. The waves would evolve by maintaining the form shown, but rotating around the basin.

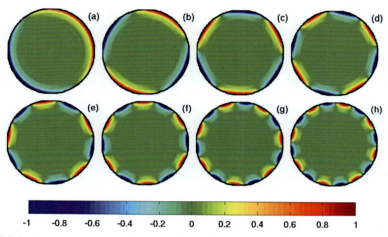

FIGURE 5.3

Numerically computed real part of the vertical mode-1 displacement, $G(r)$, for the first six Kelvin modes of a model large circular mid-latitude lake.

5.6 The circular lake: discussion

As we did with the discussion of the channel, let's take a bit of space to discuss why the mathematics of this problem is a bit exotic. One might start by noting that (5.47) has terms which are potentially problematic at, or near, $r = 0$ because they have inverse powers of r. For the analytical theory uses the Bessel functions, this is actually a good thing because it allows for the choice of solution in terms of the "correct" Bessel function (or modified Bessel function for Kelvin waves). The other Bessel function (or modified Bessel function for Kelvin waves) is simply thrown out. Were the Bessel and modified Bessel functions to be actual closed form functions this might be worth pursuing this route in more detail. However, Bessel and modified Bessel functions are so-called special functions, meaning they are an infinite series defined as the solution to (5.47). They are thus convenient to write down, with a large amount of mathematical theory dating back to the age that precedes the availability of computers, but often a bit tricky to work with computationally. Most modern mathematical software (Maple, Mathematica, MATLAB®, sci py) have routines for their evaluation. But for a computationally minded lake scientist, this seems like the ratio of mathematics to utility is not favorable!

Indeed, it is unclear whether using the Bessel and modified Bessel equations is the most efficient way to present solutions, in the modern context. This is because numerical methods for a general basin have been derived, and for most applications the solution must be evaluated numerically anyway (and even then it still needs to be presented visually).

82 CHAPTER 5 Rotating shallow water dynamics

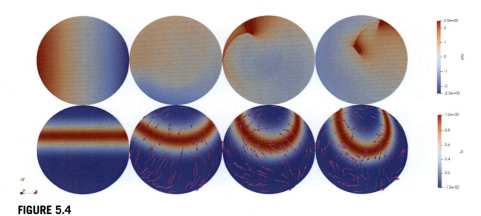

FIGURE 5.4

Top: Snapshots of the interface displacement η. **Bottom:** Snapshots of the passive tracer B with velocity direction field super-imposed (purple arrows).

5.7 The tilted free surface problem

With a bit of theory under our belt, let's turn to an actual solution of the full equations in a model lake. We will stick to the simplest basin, namely a round lake, as well as very simple initial conditions: the free surface is assumed to have a linear tilt from west to east (x) and no initial variation in the north-south direction (y), and both velocity components are assumed to vanish. To visualize what the motion in the lake actually accomplishes we place a passive tracer in the lake that is moved by the velocity field,

$$B_t + \nabla \cdot (B\vec{u}) = 0 \qquad (5.51)$$

The passive tracer is in itself an idealization, since all tracers in an actual lake would be mixed as well as advected. Biological tracers, like zoo and phytoplankton, would also evolve (e.g. zooplankton would eat phytoplankton). Nevertheless, starting with a simple tracer makes interpretation of results easiest. After some trial and error, which we do not show, we have chosen to initialize the passive tracer as a strip running west to east, just to the north of the mid-point of the lake. This will allow us to clearly see the effects of rotation.

The results of the simulation are shown in Fig. 5.4. The upper row shows the evolution of the free surface, while the lower row shows the evolution of the tracer. The upper panel clearly demonstrates that the initially linear tilt in the free surface rapidly changes in shape. One could say it leads to the generation of a number of different waves, all modified to some extent by the Earth's rotation. In terms of wave-type, both Kelvin-wave like, coastally trapped disturbances that rotate around the basin and propagating Poincaré waves that move back and forth across the basin can be seen. Moreover, effects of nonlinearity lead to the formation of sharp fronts, in the two rightmost panels of the upper row. These fronts are completely outside of the purview of the linear theory discussed earlier.

The dynamics of the passive tracer, shown in the lower panel are considerably more sensitive to currents that maintain a fixed geometric structure for longer periods of time. In this context this is provided by the Kelvin waves. However, even here there is an unexpected feature in that both the easternmost (right) and westernmost (left) portions of the passive tracer distribution are deflected to the north. Thus for later times the passive tracer forms a horseshoe shaped pattern.

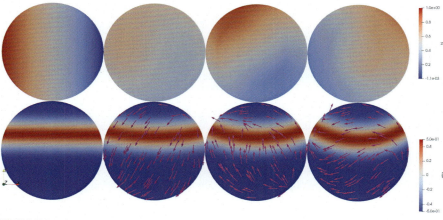

FIGURE 5.5

A simulation with an initial free surface tilt that is one fifth of that shown in Fig. 5.4. **Top:** Snapshots of the interface displacement η. **Bottom**: Snapshots of the passive tracer B with velocity direction field super-imposed (purple arrows).

While we have noted that the observed pattern of tracer is largely controlled by the Kelvin modes, a reader may wonder to what extent the formation of sharp fronts influences large scale tracer transport. In Fig. 5.5 we show the results for a simulation with an initial tilt that is one fifth of the one shown in Fig. 5.4. It can be seen that the free surface does not form any sharp fronts (upper row; and note the change of colorbar between the figures). However the systematic, northward deflection of the tracer band is again observed, albeit to a far smaller degree.

While "whole basin" visualizations are nice, it is perhaps useful to get a look at profiles of the free surface as line plots. For the circular basin, it takes some thought as to the best way to extract such profiles. In Fig. 5.6 this is accomplished by plotting free surface elevation as a function of the distance from the wall. The location in the azimuthal, or around the lake, direction is chosen so that the profile passes through the largest free surface deflection. It can be seen that at no point is the profile dominated by just the decay scale associated with the Kelvin wave. Thus, the natural world does not simply select a single mathematical form, and a more complete biophysical model should not be based on mathematical idealizations at the level of linear theory.

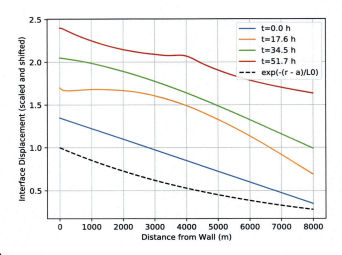

FIGURE 5.6

Radial transects of interface displacement for the 'one-fifth amplitude' linear case simulation, with varying angle $\theta > 0$ (w.r.t. the positive x-axis) chosen to pass through the maximum value of the interface displacement along the wall boundary. Different color curves correspond to time values $t = 0$ (blue), $t = 17.6$ h (orange), $t = 34.5$ h (green), $t = 51.7$ h (red). Subsequent curves have each been shifted up along the y-axis by 0.35 relative to the previous curve. The dashed line is a reference curve depicting the theoretical decay scale of an idealized Kelvin wave trapped to the wall boundary with Rossby deformation radius $L_0 = c_0/f = 6825$ m and wall boundary at $r = a$. Here, $a = 8000$ m, $f = 7.88 \times 10^{-5}$ s^{-1}, and $c_0 = 0.50$ m s^{-1} is the linear long wave speed.

One important aspect of this nonlinear problem is the clear appearance of steepening and front formation. The form taken in this problem is quite complex due to the fact that multiple wave types are possible. If we return to the non-rotating shallow water theory in one dimension we can get much more immediate intuition. The governing equations read

$$\frac{\partial u}{\partial t} + u\frac{\partial u}{\partial x} = -g\frac{\partial \eta}{\partial x} \tag{5.52}$$

$$\frac{\partial \eta}{\partial t} + \frac{\partial u(H + \eta)}{\partial x} = 0 \tag{5.53}$$

and we can get the essence of nonlinearity by assuming the free surface is zero so that the first equation reduces to

$$\frac{\partial u}{\partial t} = -u\frac{\partial u}{\partial x}.$$

This is like an advection equation except that the propagation speed is set by the value of u. In particular, if u is larger, disturbances propagate faster. If we do the thought experiment of how a Gaussian hump disturbance would evolve, we arrive at a mental picture that looks somewhat like Fig. 5.7. The initial crest is denoted by a star and has the largest propagation speed. The initial mid-height is denoted by a triangle and has smaller propagation speed than the region denote by the star. The edge of the disturbance with a nearly zero propagation speed is denoted by a circle. At a later time, the star has caught up to, and slightly surpassed, the triangle, which has in turn nearly caught up to the circle. The disturbance takes the familiar form of a surfer's wave at the point where it is starting to break.

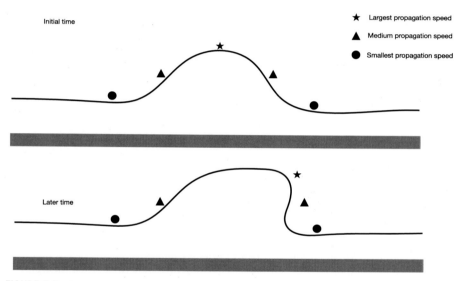

FIGURE 5.7

Conceptual picture of the effect of nonlinearity in the shallow water equations. The regions with the faster local speeds of propagation catch up to the slower speeds leading to a nearly vertical wave front, and eventually overturning.

While a complete description of nonlinear effects requires far more significant mathematics, and when rotation is included the problem has an interesting numerical side ([6]) the intuition from the above suggests sharp fronts are expected when there is little dissipation in the system. We will see that this is in fact not necessarily true, when we discuss the "missing physics" of wave dispersion in the next chapter.

5.8 An instability problem: simulation

The tilted free surface of a round basin problem, while of considerable utility in a laboratory setting, is not likely to be representative of natural phenomena. Of course, the case can be made that it is a useful cartoon of a particular lake of interest, but it

86 CHAPTER 5 Rotating shallow water dynamics

must be admitted that in natural waters it is far more likely that the mathematically tidy solution is likely to persist for only a short amount of time.

The evolution away from a particular state is called a "stability problem", and in the mathematical study of fluid mechanics, hydrodynamic stability has a long and distinguished history. This is in part due to the fact that over the years, it has allowed applied mathematicians access to new and interesting mathematics. In a computational setting, the focus is somewhat different, and in this section we present one example of a particularly simple solution evolving to a more complex state.

The solution we will consider is the so-called geostrophic double jet, or in other words, a jet that is initially in equilibrium with the effect of the Earth's rotation. The balance with the Earth's rotation implies that the free surface must be deflected in order to drive the pressure distribution so that there is flow along lines of constant pressure. In practical terms, two jets form on the flanks of a hump of fluid. The domain is a channel with periodic ends, so that the double jet is free to evolve as per the governing equations. We give this evolution a significant kick, by introducing a peninsula to the upper wall of the channel.

Because wind-driven jets are rarely observed to reach to the bottom of natural bodies of water, we use a reduced gravity model, with an active layer thickness $H_0 = 7.5$ m, a mid latitude value of $f = 7.8825 \times 10^{-5}$ s^{-1}, an initial amplitude of the "hump" that is 6.5% of the layer depth, and a functional form of the free surface height

$$\eta = a \exp\left[-\left(\frac{y-L}{W}\right)^2\right].$$

The numerical model employs a Discontinuous Galerkin finite element method. This method will be discussed in Chapters 10 and 11. Numerical experiments with standard triangle based grids yielded some spectacular failures near the peninsula. The peninsula thus proved the need for the curved elements/cubature rules discussed in Chapter 11, and the reader is directed there for a detailed discussion. Dissipation in the model was provided by a standard quadratic stress law

$$c_D(x, y)\vec{u}|\vec{u}|,$$

where the damping coefficient was increased on phenomenological grounds near the solid boundaries. The coefficient c_D was varied such that it decays linearly to 0 in the interior of the domain from a maximum value of 0.0025 in a damping region near the boundary of width 125 m.

In Fig. 5.8 we show the evolution of the horizontal velocity, or in other words of the double jet. The four panels show the velocity field at $t = 0$, 1.88 h, 3.76 h, 5.64 h, with a blue-red colorbar saturated at ± 0.26 ms^{-1}. It can be seen that the double jet is modified by the peninsula, and degenerates into a number of vortices that eventually fill the domain.

In Fig. 5.9 we show the evolution of a tracer that is initially distributed in a Gaussian-shaped region (a round blob) near the peninsula. It can be seen that early

5.8 An instability problem: simulation

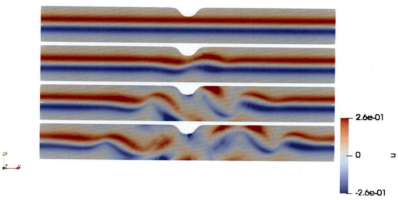

FIGURE 5.8

Snapshots of the u field for the breakdown of a double-jet at times (from top to bottom) $t = 0$, 1.88 h, 3.76 h, 5.64 h.

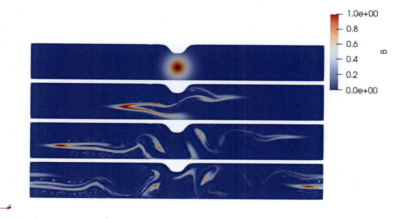

FIGURE 5.9

Snapshots of the B (tracer) field for the breakdown of a double-jet at times (from top to bottom) $t = 0$, 1.88 h, 3.76 h, 5.64 h.

evolution consists of advection to the right and left by the twin jets, but in the third panel some tracer has been caught up in the growing vortices. By the final panel the tracer has very different patterns in the region near the peninsula and well away from it.

As with any single case simulation, it is important to ask "how generic is the result"? To help answer this question we halved the initial velocity and repeated the simulation. The results of this simulation are shown in Figs. 5.10 and 5.11 for the u component of velocity and tracer concentration, respectively. It can be seen that at a qualitative level the observations about the peninsula inducing an instability of

88   CHAPTER 5 Rotating shallow water dynamics

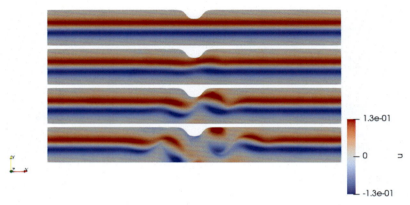

FIGURE 5.10

Same as Fig. 5.8 but the jet velocity has been reduced by a factor of a half.

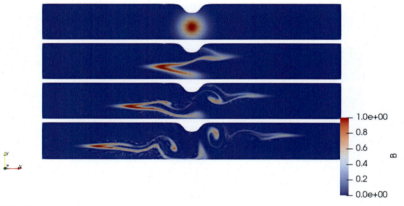

FIGURE 5.11

Same as Fig. 5.9 but the jet velocity has been reduced by a factor of a half.

the jet remain valid. The instability is less energetic, and hence spreads away from the peninsula more slowly. Indeed, in the u field the double jet remains coherent well away from the peninsula. The tracer transport has clear features associated with the double jet, and the vortices that do form remain coherent, on the upstream and downstream flanks of the peninsula.

Both the tilted free surface and the unstable jet simulations are similar, in the sense that the initial state is given a somewhat arbitrary form, and then the flow is allowed to evolve according to Newton's laws. This may prove useful in isolating certain key phenomena a researcher is interested in. A modeling effort that seeks to reproduce data would have to be structured rather differently: the model lake would have to be brought into a state that is similar to that in the real world by a combination of forcing (mainly wind, but possibly also radiation) and dissipation. In more complete models (e.g. the MIT gcm, ROMS) the attention paid to the manner in which the forcing and dissipation are carried out often almost matches the dynamics itself, and the many parameters needed prove quite difficult to constrain ([3]).

A historically minded reader will note that past practice in environmental fluid mechanics has attached more importance to those "process studies" that were closely linked to analytical solutions. This was largely due to the lack of available computer resources and analysis tools. Given that both are available in abundance, there is no clear reason to prefer the tilted round lake to the unstable jet problem, just because the former is tied to a classical solution ([2]).

5.9 Mini-projects

1. Find the eigenfunctions of the matrix (5.5) and use these to discuss how the various waves are expressed in the u, v, and η fields.
2. Reconsider the problem of plane waves of the shallow water system. Instead of using linear algebra derive a single equation for the v variable, and then assume the wave ansatz for v. Show that the dispersion relation is unchanged.
3. Recall that the dispersion relation for Poincaré waves shows that the phase speed can grow without bound. Find examples of other physical systems for which this may occur and discuss their physics.
4. Discuss the properties of the Bessel and modified Bessel functions (e.g. see [8] for a nice historical discussion) and what they imply for the modes in a circular domain.
5. Using a reference of your choice (e.g. [5]) discuss the stability of a shear flow in a non-rotating system. How are the results modified on the f-plane (this is often referred to as **barotropic instability**.

References

[1] Leon Boegman, et al., High-frequency internal waves in large stratified lakes, Limnology and Oceanography 48 (2) (2003) 895–919.
[2] G.T. Csanady, Large-scale motion in the Great Lakes, Journal of Geophysical Research 72 (16) (1967) 41514162.
[3] G. Djoumna, K.G. Lamb, Yerubandi R. Rao, Sensitivity of the parameterizations of vertical mixing and radiative heat fluxes on the seasonal evolution of the thermal structure of Lake Erie, Atmosphere-Ocean 52 (4) (2014) 294–313.
[4] A. Gill, Atmosphere-Ocean Dynamics, 1st ed., Academic Press, 1982.
[5] P.K. Kundu, I.M. Cohen, Fluid Mechanics, 4th ed., Elsevier Academic Press, 2008.
[6] Allen C. Kuo, Lorenzo M. Polvani, Time-dependent fully nonlinear geostrophic adjustment, Journal of Physical Oceanography 27 (8) (1997) 1614–1634.
[7] Martina Preusse, Heinrich Freistuhler, Frank Peeters, Seasonal variation of solitary wave properties in Lake Constance, Journal of Geophysical Research: Oceans 117.C4 (2012).

90 CHAPTER 5 Rotating shallow water dynamics

[8] George F. Simmons, Differential Equations With Applications and Historical Notes, CRC Press, 2016.

[9] R. Stocker, J. Imberger, Energy partitioning and horizontal dispersion in a stratified rotating lake, Journal of Physical Oceanography 33 (2003) 512–529.

CHAPTER

Rotating shallow water dynamics: Dispersion and nonlinearity

6

CONTENTS

6.1 Realism in wave models.. 91
6.2 The dispersive shallow water model 92
6.3 Adjustment problems on the field scale 96
6.4 Seiche evolution.. 102
6.5 Coding ideas 110
6.6 Mini-projects 114
References... 114

6.1 Realism in wave models

From the point of view of a practitioner the previous chapter is quite disheartening. Much of the classical theory seems to be the tail wagging the dog, with mathematical descriptions driving both choices of methodology and the nature of conclusions once can draw. The classical "Ockham's razor" point of view (i.e. build the simplest possible model that is consistent with the data) is thus challenged by the desire to find conclusions as formulae (i.e. in closed form).

On the other hand, the numerical methods featured in the previous chapter are quite advanced, and beyond the set of tools that a typical scientist has experience with. We will thus take a step back and work to close the gap between theory and model. We will do this by examining the physics (and indeed the missing physics) of the shallow water model, and will build more complex simulations from very simple, laboratory inspired, set ups. This will be developed over the space (Chapters 6–8).

In this chapter we develop simple models that we have no intention of solving analytically. We choose these models so that they extend the classical single layer shallow water theory so that the model is more representative of what is already known about real (internal) waves in lakes:

1. Real waves are dispersive (i.e. waves of different length scales travel at different speed).
2. Real waves are finite amplitude (i.e. the nonlinear terms in the governing equations should not be summarily dropped).
3. Real waves are affected by the Earth's rotation.

Physics and Ecology in Fluids. https://doi.org/10.1016/B978-0-32-391244-0.00016-4
Copyright © 2023 Elsevier Inc. All rights reserved.

4. Real waves are affected by small scale turbulence and viscosity.

For the first three points tidy model formulations are available. The final point remains a somewhat open question and a satisfying answer will not be shown to the reader until Chapters 7 and 8.

The role theory plays in this chapter is to set the stage for numerical methods. In particular, we want all assumptions to be clearly laid out. We of course assume that the numerical solution methods of the models we show are reasonable to derive, code, and run. More often than not, in the modern computing landscape, this is a good assumption and the pathologies of numerical analysis rarely show up in practice.

6.2 The dispersive shallow water model

The classical shallow water equations without rotation are a common feature of both advanced numerical methods and advanced partial differential equations courses. This is because they are an example of a nonlinear hyperbolic system ([4]). They thus carry information exactly along so-called characteristics curves, and for a very wide class of initial conditions they form shocks in finite time.

Shocksare solutions with a jump, or discontinuity, basically a mathematical idealization of the "surfer's" wave shown in Fig. 5.7. Due to their discontinuous nature, shocks appear to violate the idea of an equation involving spatial derivatives, and the mathematics of describing shocks (i.e. weak solutions) is thus quite impressive. Breaking waves do form in lakes, but they do so at scales of description far shorter than those of a typical lake model's single grid box. As such it seems a strange area of focus for a practically minded modeler (an excellent, if sophisticated introduction can be found in [5]).

A different topic of classical applied mathematics proves far more profitable. It has been known since the 19th century that the Korteweg-de Vries (or KdV) allows for so-called solitary wave solutions, or waves that solve a nonlinear equation but do not change shape as they propagate ([8]). The KdV equation is a single-direction advection equation augmented by both nonlinear and dispersion terms,

$$\frac{\partial A}{\partial t} = -c\frac{\partial A}{\partial x} + \alpha A\frac{\partial A}{\partial x} + \beta\frac{\partial^3 A}{\partial x^3}$$

where α and β are physical parameters determined from local conditions (i.e. the stratification in a particular lake). In the second half of the 20th century an amazing amount of applied mathematics was developed for this equation; the so-called inverse scattering, or soliton, theory.

The primary lesson from the KdV equation for our modeling purposes is that if dispersion is present in a system it may balance nonlinearity and preclude shock formation. Both surface water waves and internal waves are well known to exhibit dispersion (with longer waves observed to travel faster), and hence a model with

6.2 The dispersive shallow water model

dispersion is highly desirable. Indeed, the classical shallow water equations without rotation are a very rare example in the physical description of waves that yields nondispersive waves. The classical shallow water waves have many attractive features, since they represent the typically small aspect ratio of natural motions in lakes (i.e. horizontal length scales that are much larger than vertical length scales). Thus an attractive applied mathematical problem is "How can we most economically represent dispersion in extensions of the classical SW system".

The underlying idea for representing wave dispersion in layered model equations is that weak vertical accelerations (which lead to nonhydrostatic pressure) should be parametrized as horizontal dispersion. The route to this parametrization can be quite mathematically laborious, but the key idea is relatively simple. We follow the work of de la Fuente et al. [2], who studied internal waves in a circular basin for a single fluid layer. The governing equations used by these authors read

$$\frac{\partial h}{\partial t} + \nabla \cdot (h\mathbf{u}) = 0, \tag{6.1}$$

$$\frac{\partial (uh)}{\partial t} + \nabla \cdot ((uh)\mathbf{u}) = -gh\frac{\partial \eta}{\partial x} + fvh + \frac{H^2}{6}\frac{\partial}{\partial x}\left(\nabla \cdot \frac{\partial (uh)}{\partial t}\right), \tag{6.2}$$

$$\frac{\partial (vh)}{\partial t} + \nabla \cdot ((vh)\mathbf{u}) = -gh\frac{\partial \eta}{\partial y} - fuh + \frac{H^2}{6}\frac{\partial}{\partial y}\left(\nabla \cdot \frac{\partial (uh)}{\partial t}\right), \tag{6.3}$$

where $\mathbf{u} = (u(x, y, t), v(x, y, t))$ is the velocity field, $h(x, y, t) = H(x, y) + \eta(x, y, t)$ is the total depth with H representing the undisturbed depth, and η is the free surface displacement. The constants g and f are the acceleration due to gravity and the Coriolis frequency, respectively. The difference between the set of equations (6.1)–(6.3) and the traditional shallow water model is the inclusion of the dispersive terms $\frac{H^2}{6}\nabla(\nabla \cdot (\mathbf{u}h)_t)$ found in the momentum equations (6.2) and (6.3). Indeed, the above system was first proposed by Brandt et al. [1] in their study of internal waves in the Strait of Messina. As mentioned above, this system is derived by a perturbation expansion in powers of the small dimensionless parameter $\mu = (H/L)$, and therefore is only physically relevant if $\mu \ll 1$ ([2]). As a small aside, note that the variable we refer to as h, is referred to as z by some authors. We found this notation to be too confusing when comparing various references, and perhaps by noting it here we can save the reader some time.

In the above form of the equations, we have neglected bottom and surface stresses in Eqs. (6.1)–(6.3) since their inclusion into numerical schemes is conceptually simple and contributes little to the discussion. We will use the quadratic stress in this chapter which would take the form

$$-c_d |\vec{u}| \vec{u} h \tag{6.4}$$

where the factor of h is included because the equations above are given for $\vec{u}h$ and not $\vec{u}$.

94   CHAPTER 6   Rotating shallow water dynamics

To relate these equations to the classical shallow water equations, consider the case of one dimensional flow without rotation, with a flat bottom. Some simple algebraic manipulations lead to the set of equations,

$$u_t + uu_x = -g\eta_x + \frac{H^2}{6} u_{xxt} \tag{6.5}$$

$$\eta_t + [(H + \eta)u]_x = 0. \tag{6.6}$$

Writing the equations this way we can see that the conservation of mass equation is unchanged, but the conservation of momentum has an additional term. This term represents the nonhydrostatic pressure changes, but manifests as a mixed space-time derivative of the velocity u. The challenge for numerical methods is how to integrate this new term into the time-stepping method. Some methods, such as those used later in this chapter have a very easy time with this. For others, like those discussed in Chapters 9–11, it can be a challenge.

The linearized form of Eqs. (6.1)–(6.3) in one space dimension with $f = 0$ take the form

$$\eta_t + (Hu)_x = 0, \tag{6.7}$$

$$u_t = -g\eta_x + \frac{H^2}{6} u_{xxt}. \tag{6.8}$$

If we again assume a flat bottom ($H =$ constant) and make the wave ansatz, or that η and u are proportional to $e^{i(kx-\sigma t)}$, we can derive the dispersion relation

$$\sigma^2 = \frac{gHk^2}{1 + \frac{H^2}{6}k^2}, \tag{6.9}$$

with phase speed

$$c_p = \frac{\sigma}{k} = \pm \frac{\sqrt{gH}}{\left(1 + \frac{H^2}{6}(k^2 + l^2)\right)^{1/2}}. \tag{6.10}$$

The shallow water speed is recovered in the long wave limit as $k \to 0$,

$$c_p = \frac{\sigma}{k} \to \pm\sqrt{gH}. \tag{6.11}$$

The phase speed decreases from this value and does so significantly when k is large (short waves). It is thus clear what this new term can get us; as wave steepening due to nonlinearity begins to play a role, we effectively activate waves with larger and larger values of k, but the dispersion then causes these waves to lag behind, giving some hope that for many cases shock formation will not play a prominent role in the dynamics. A discussion of the group speed is left to the reader as an exercise.

6.2 The dispersive shallow water model

For the interested reader, the above expressions (6.9) and (6.10) can be compared to the exact dispersion relation for gravity waves in a single layer fluid derived from potential flow theory ([3]):

$$c_p = \frac{\sigma}{k} = \pm\sqrt{\frac{g\tanh(kH)}{k}}. \qquad (6.12)$$

The phase speeds (6.11), (6.12), and (6.12) are compared in Fig. 6.1. All three dispersion relations agree in the long wave limit $k \to 0$, and the dispersive shallow water model's dispersion relation agrees qualitatively with the behavior of the full dispersion relation for gravity waves (6.12).

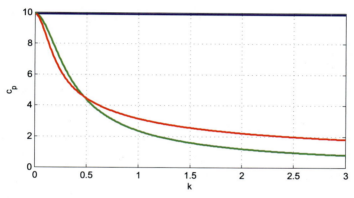

FIGURE 6.1

Comparison of phase speeds from the traditional shallow water mode (blue), the dispersive shallow water model (green), and the full dispersion relation from potential flow theory (red).

The importance of the dispersion is made clear in a standard "adjustment" simulation. The relevant code is given in

`swnh1d_shortscale.m`

with parameter values listed in Table 6.1. The domain is periodic in x for simplicity, extending over $-L \leq x \le L$. An initial increase in the free surface height occupying $-0.1L \leq x \le 0.1L$, with flanks smoothed over a region $0.01L$ thick, is allowed to evolve. This yields both rightward and leftward propagating waveforms. For the rightward propagating waveforms, the resulting free surface (scaled by the total depth H) and horizontal velocity (scaled by the wave speed c_0) are shown in Figs. 6.2 and 6.3 for early and late times, respectively. It can be seen that even at early times, the classical SW solution (in red) steepens to form a shock, which is then somewhat smoothed by the numerical method (see the discussion in the final section of this chapter for details). In contrast the equations with the dispersive correction lead to the formation of a wave train and no sign of a shock is observed. Indeed, for later times the train of waves is quite clear, with the largest wave leading (i.e. the wavetrain is rank-ordered).

96 CHAPTER 6 Rotating shallow water dynamics

Table 6.1 Parameters for the Adjustment case.

Label	N	L (m)	H (m)	g (m s^{-1})	η_0 (m)	f (s^{-1})
Short adjustment	2048	1000	10	0.0981	2	0

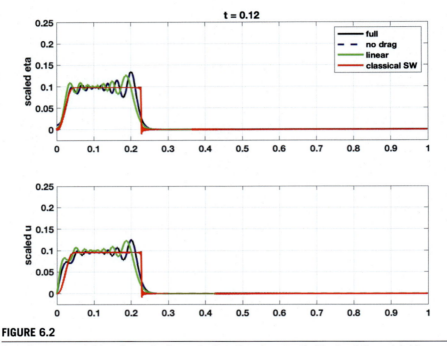

FIGURE 6.2

An adjustment problem for early dimensionless time ($t = 0.12$). Black – full model, blue – model without drag, green – linearized model, red – classical SW (no dispersion). Upper panel – free surface scaled by undisturbed depth, Lower – panel u scaled by c_0.

For the times shown the quadratic bottom stress damping has little effect. For completeness we show the evolution of the linearized system in green. It can be seen that linearized system leads to a breakdown into an incoherent wave field (due to dispersion). This is due to the fact that steepening cannot occur (because the nonlinear terms have been dropped), and hence dispersion cannot be balanced, as in the classical KdV equation.

6.3 Adjustment problems on the field scale

The sample numerical simulation discussed in the previous section was tailored to illustrate the effects of nonlinearity and dispersion. It was thus a little on the unreal-

6.3 Adjustment problems on the field scale

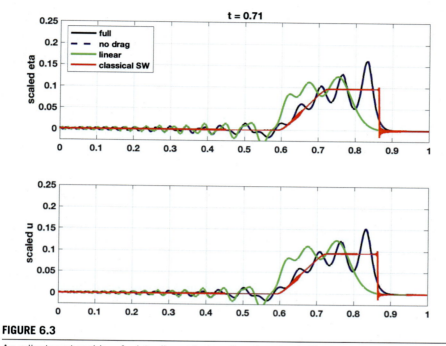

FIGURE 6.3

An adjustment problem for late dimensionless time ($t = 0.71$). Black – full model, blue – model without drag, green – linearized model, red – classical SW (no dispersion). Upper panel – free surface scaled by undisturbed depth, Lower – panel u scaled by c_0.

istic side. In particular, the initial perturbation was rather large in amplitude (20% of the undisturbed depth) and the simulation was not run for a particularly long time.

Table 6.2 Parameters for the Field Scale Adjustment cases.

Label	N	L (m)	H (m)	g (m s^{-1})	η_0 (m)	f (s^{-1})
Long adjustment	65,536	2×10^5	10	0.0981	1	1×10^{-4}

In this section we discuss a suite of simulations that are tailored to the question "what would we actually expect to happen on the field scale". We repeat the set up from the previous section, but in a much larger domain (200 km for the rightward propagating wavetrain we concentrate on). We also decrease the amplitude of the initial perturbation to 10% of the undisturbed depth, and increase the extent of the smoothed region on the flanks to 2 km. See Table 6.2 for a complete list of parameter values.

We define a characteristic time as the time a linear wave would take to traverse half of the domain. This yields a value of just over 56 hours, implying that rotation may play a significant role in the dynamics.

98 CHAPTER 6 Rotating shallow water dynamics

We use a very large number of grid points, in order to successfully resolve all possible phenomena that our model equations represent: propagation, dispersion, nonlinearity, rotation, and damping by bottom stress. The horizontal grid resolution is just over 6 meters. While this is certainly a much finer resolution than many published lake modeling studies (we will discuss the reasons for this below), it is not a particularly difficult task for even a modern lap top. The relevant code is given in

```
swnh1d_large_adjust.m
```

with paremeters listed in Table 6.2.

Before we turn to the results of the numerical experiments it is useful to think about what information we could imagine gaining by running them. The easiest answer is that we will be able to conduct pairwise comparisons between cases with and without a particular physical mechanism (for example with and without the effect of the Earth's rotation). These differences could be quantitative in nature, but the hope is that if the differences are truly important, they will lead to qualitative differences. Since wave motions are likely part of what we will see, we can imagine comparing propagation speeds of any waves we see, and their shape. We will plot all three fields in the equations, the free surface height, η scaled by the undisturbed depth, and the two components of velocity u and v (the latter only activated in the presence of the Earth's rotation) both scaled by the shallow water wave speed $c_0 = \sqrt{gH}$. We chose to present a case of internal waves where the density jump across the light fluid-heavy fluid interface is one percent of the reference density.

Figs. 6.4, 6.5, and 6.6 show the state of the simulation for early, medium, and late times (measured by how much of the subdomain shown the waves have propagated across).

For early times it is already clear that the dominant qualitative difference is between cases with rotation (all colors but red) and without rotation (red). The second most obvious qualitative distinction is between the linearized equation case (green) and all other cases. All nonlinear cases have what appears to be a region of rapid oscillations. Indeed, the transition is so rapid, it looks like a thick pen had been used to draw the transition between the wave front and the undisturbed fluid ahead of it. With rotation, the front is followed by a long wave form extending from the origin to near $x = 0.34$. The u and v components of the velocity are out of phase for the long, rotation induced wave behind the leading transition. On the time scale of this figure, the case without bottom drag leads to waves with only a slightly larger amplitude.

Fig. 6.5 follows the evolution of the system, and while the qualitative distinctions from the above paragraph remain, there are two particularly noteworthy phenomena due to rotation that merit comment. The first of these is the large amplitude, long length scale free surface deflection near between $x = 0$ and about $x = 0.2$. This deflection is evident in both the nonlinear and linear systems (and on the scale of the image appears unaffected by bottom drag), and has the distinguishing feature that the u component of the velocity field is nearly zero, while the v component is large. This can be understood from the x component of geostrophic balance (5.3), repeated here for the reader's convenience:

6.3 Adjustment problems on the field scale   99

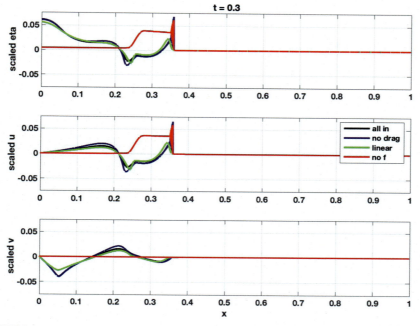

FIGURE 6.4

The Large domain adjustment problem for early dimensionless time ($t = 0.3$). Black – full model, blue – model without drag, green – linearized model, red – no rotation ($f = 0$). Upper panel – free surface scaled by undisturbed depth, Lower – panel u scaled by c_0.

$$-fv = -g\frac{\partial \eta}{\partial x}.$$

It is clear that the v component of velocity is largely set by the rate of change of the free surface height with x. This is a good example how even a limited theory, like geostrophic balance, can provide useful insight.

The second phenomenon that is clear from the figure is that the no bottom drag case has evolved into a situation with a secondary wave train (just to the right of $x = 0.5$). This feature has a clear expression in the u field, but not in the v field. It is also not clearly observed in the case with bottom drag (black curve). This is the first illustration we have of the somewhat dubious effects of bottom drag.

Fig. 6.6 shows the evolving wave trains just as they near the right boundary of the domain. The geostrophic state near $x = 0$ remains largely unchanged for both the linearized and full equations. The leading and secondary wave train are also clearly evident, though at the scale of the half-domain little detail can be made out.

Figs. 6.7 and 6.8 focus on the leading and second wave train, respectively. It can be seen that for the leading wave train the case without bottom drag leads to a wave

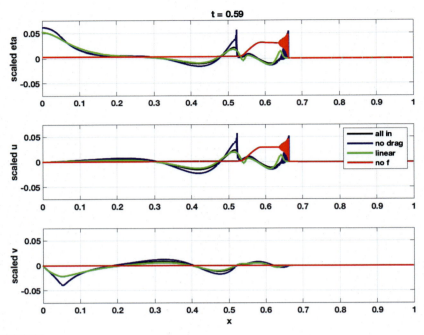

FIGURE 6.5

The Large domain adjustment problem for medium dimensionless time ($t = 0.59$). Black – full model, blue – model without drag, green – linearized model, red – no rotation ($f = 0$). Upper panel – free surface scaled by undisturbed depth, Lower – panel u scaled by c_0.

train that is ahead of that for both the rotating and non-rotating cases with drag. The non-rotating case (red) corresponds to a classical undular bore with the free surface transitioning from its undisturbed value of zero, to a non-zero downstream value after passing through a long wave train. The rotating wave train shows a slightly longer inter-crest spacing and there is a clear modulation due to the long Poincaré waves that are difficult to see on a detailed plot like Fig. 6.7. The secondary wave train shown in Fig. 6.8 is perhaps even clearer in terms of story line. The most obvious point being that there is no secondary wave train in the non-rotating case, implying that its existence is 100% due to rotation. The role of bottom drag is also clear, since the black curve shows a much smaller amplitude wave train, and one that lags some way behind the blue (no bottom drag) wavetrain.

This is an impressive amount of information for a fairly straightforward simulation. In summary, it shows that for large lakes rotation leads both to large length scale motions, as well shorter length scale motions as seen in the secondary wave train. The possible implications for transport are also diverse:

6.3 Adjustment problems on the field scale

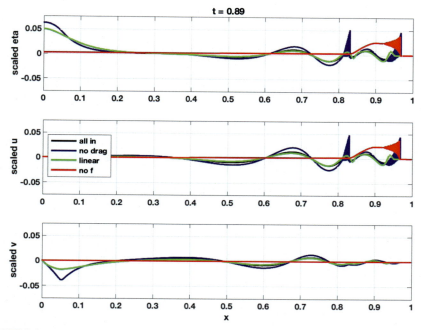

FIGURE 6.6

The Large domain adjustment problem for late dimensionless time ($t = 0.89$). Black – full model, blue – model without drag, green – linearized model, red – no rotation ($f = 0$). Upper panel – free surface scaled by undisturbed depth, Lower – panel u scaled by c_0.

- The long waves, and the geostrophic state near $x = 0$ both induce transverse (i.e. v) currents which could transport tracers in a direction that is perpendicular to the primary direction of wave motion.
- The presence of dispersion modifies the expected behavior on small length scales, leading to long lived wave trains as opposed to dissipative shocks that might disappear before reaching a significant distance in x.
- The secondary wavetrains are affected by bottom drag, but could lead to a secondary "pulse" of transport in the x direction.

We took considerable care in making sure the simulations presented had sufficient resolution, and because the simple choice of a periodic domain allowed us to use FFT-based spectral methods, there was very little numerical dissipation. This is very different from many "off the shelf" lake models and so the primary purpose of this section is to encourage readers to consider what mechanisms may matter for their chosen problem PRIOR to choosing a computational tool. A poor choice may *a priori* filter out the very phenomenon that is responsible for transport!

CHAPTER 6 Rotating shallow water dynamics

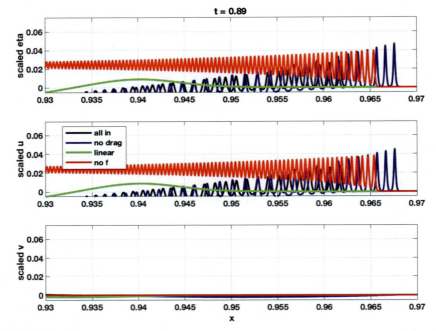

FIGURE 6.7

Detail of the Large domain adjustment problem for late dimensionless time ($t = 0.89$), focusing on the leading wavetrain. Black – full model, blue – model without drag, green – linearized model, red – no rotation ($f = 0$). Upper panel – free surface scaled by undisturbed depth, Lower – panel u scaled by c_0.

6.4 Seiche evolution

While the initial conditions chosen in the last section are a standard tool of numerical methods, outside of engineering applications in which an actual lock in a canal is suddenly opened, they are some ways away from what typically occurs in the natural world.

The question of coming up with a simple, yet applicable set of initial conditions is actually quite difficult. Nevertheless, the observation that mid-to-large natural bodies of water experience oscillations called seiches has led to a considerable history of studying periodic, or nearly periodic, motions in lakes.

Table 6.3 Parameters for the Field Scale Adjustment cases.

Label	N	L (m)	H (m)	g (m s^{-1})	η_0 (m)	f (s^{-1})
Seiche breakdown	8192	1×10^4	10	0.0981	0.5	1×10^{-4}

6.4 Seiche evolution

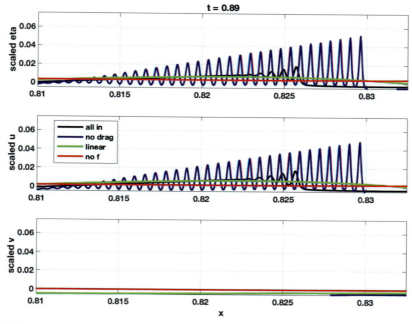

FIGURE 6.8

Detail of the Large domain adjustment problem for late dimensionless time ($t = 0.89$), focusing on the secondary wavetrain. Black – full model, blue – model without drag, green – linearized model, red – no rotation ($f = 0$). Upper panel – free surface scaled by undisturbed depth, Lower – panel u scaled by c_0.

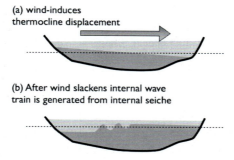

FIGURE 6.9

Schematic of wind-induced changes in a stratified lake, panel (a); and the development of internal wave trains when the wind slackens, panel (b).

In nature, it is believed that seiching behavior most often occurs when a sustained wind pushes water from the upwind to the downwind side of a lake (see the schematic

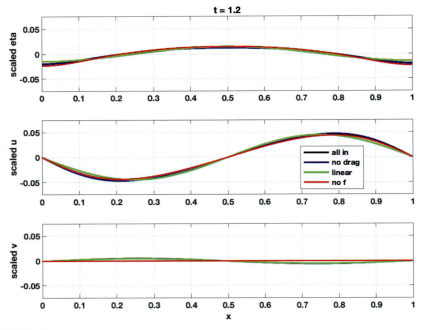

FIGURE 6.10

The seiche development problem for early dimensionless time ($t = 1.2$). Black – full model, blue – model without drag, green – linearized model, red – no rotation ($f = 0$). Upper panel – free surface scaled by undisturbed depth, Lower – panel u scaled by c_0.

diagram in Fig. 6.9). This leads to an increase in water depth on the downwind side, and crucially, pushes the interface between warm and cold water (i.e. the thermocline) down. When the wind slackens a seiche is set up, with the internal seiche in the interior (where the density change across the thermocline is much smaller) far more prominent. Surface, or barotropic seiches can be important as well, and indeed are standard cottage owner lore along the beaches of the Laurentian Great Lakes in Canada. We will concentrate on internal seiches in this section.

We choose a domain that is large enough to be affected by rotation, but not as large as the lock release of the previous section. Our computational domain occupies $-10 < x < 10$ km, and we will concentrate the figures on $0 < x < 10$ km only. The internal seiche is defined to be a simple cosine. More realistic options will be discussed at the end of this section. The amplitude is taken as 5% of the total depth, which is itself taken to be constant (effectively treating the lake as an aquarium; again a mathematical convenience). The relevant code is given in

```
swnh1d_seiche.m
```

with parameters listed in Table 6.3.

6.4 Seiche evolution

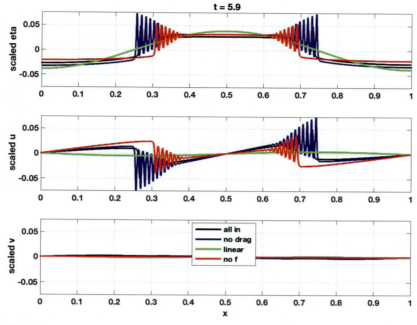

FIGURE 6.11

The seiche development problem for early dimensionless time ($t = 5.9$). Black – full model, blue – model without drag, green – linearized model, red – no rotation ($f = 0$). Upper panel – free surface scaled by undisturbed depth, Lower – panel u scaled by c_0.

Figs. 6.10, 6.11, and 6.12 show the seiche development at three qualitatively different stages of development. The same convention as the previous section is used: the free surface, and both components of velocity are shown, scaled by the domain half-width and linear shallow water speed, respectively. Time is scaled by the advective time scale, or the time it would take a linear shallow water wave to propagate across the domain half width (i.e. $T_{adv} = L/c_0$).

Fig. 6.10 shows that all four numerical experiments behave quite similarly for early times. This means that for a study that only considers short times, it may well be that the bother of nonlinearity in particular, can be safely ignored.

However, Fig. 6.11 serves as an immediate cautionary tale. Here the linearized case (green curve) retains its sinusoidal form, while all the other cases take on a profoundly different shape. This shape is dominated by a wavetrain of rank-ordered "humps" similar to what we saw in the adjustment problem of the previous section. Careful examination shows that the cases with rotation (blue and black curves) lead to a train that is faster than the case without rotation (red curve). Moreover the inclusion of a quadratic bottom drag (black curve) leads to a smaller amplitude wavetrain, and one that lags behind the drag free case (blue curve).

CHAPTER 6 Rotating shallow water dynamics

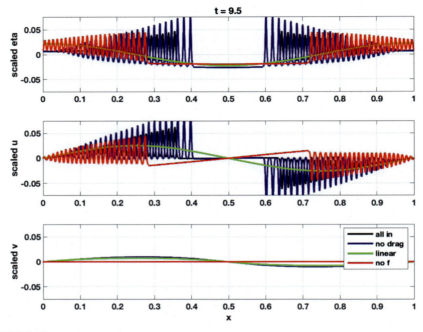

FIGURE 6.12

The seiche development problem for early dimensionless time ($t = 9.5$). Black – full model, blue – model without drag, green - linearized model, red – no rotation ($f = 0$). Upper panel – free surface scaled by undisturbed depth, Lower – panel u scaled by c_0.

Fig. 6.12 shows that these wave trains fill larger and larger portions of the domain as time increases. The role of bottom drag at keeping amplitudes modest is quite clear in this figure. A reasonable question for the modeler is whether a drag term of the form $-c_d \vec{u}|\vec{u}|$ is sufficient to account for all the possible dissipation mechanisms in a real lake. This is a bit of a strawman argument, with "No" being the obvious answer. The discussion is somewhat irrelevant to most publicly available lake models, since these adopt eddy viscosity and diffusivity values so large that most features shown in our simulations would be smoothed out quickly. We will return to this discussion when we present the finite volume and finite element methods that are typically used by lake models in Chapters 9–11 of this book.

The previously discussed three figure set is useful for getting the broad strokes idea of how the seiche evolves. A few of the details merit some further discussion. In the very early stages, all four of the model types gave very similar results. While it is clear that eventually the four models are quantitatively different, one clear question is whether differences occur before or after the wavetrains form (except for the linear case which does not form wavetrains). Fig. 6.13 shows the four cases at dimensionless time $t = 3.6$. It is clear that wavetrains have not formed at this time, but all three

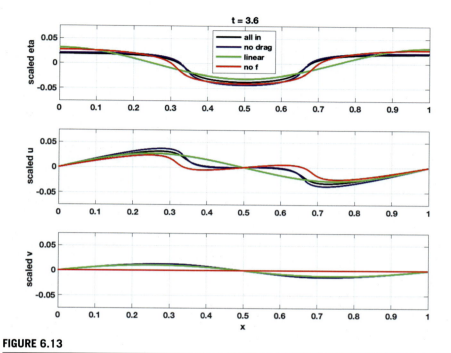

FIGURE 6.13

The seiche development problem showing differences prior to wavetrain formation; at dimensionless time $t = 3.6$. Black – full model, blue – model without drag, green – linearized model, red – no rotation ($f = 0$). Upper panel – free surface scaled by undisturbed depth, Lower – panel u scaled by c_0.

of the nonlinear cases have led to a steepening of the initially sinusoidal waveform. The clearest difference is between the rotating (black and blue) and non-rotating case (red). The quadratic bottom drag (black) has led to a slight decay of the wave amplitude, perhaps clearest in the u field (middle panel).

These differences become considerably clearer when the wavetrains develop. Fig. 6.14 shows a detail of the wavetrains at $t = 5.9$. Again, the difference between the non-rotating (red) and rotating (blue and black) cases is clear. However, the effect of the quadratic damping is now pronounced with the leading crest of the wave train for the black curve reduced in amplitude by about 30% compared to the blue curve.

The seiche breakdown results provide some important takeaways as we build up to more complex cases. First of all, they reaffirm the need for resolution from the previous section. Second of all, they point to the ubiquity of nonlinear wave trains. In the case of seiches these fill the entire domain. This is partly due to the extreme simplicity of the 1D domain, and indeed for a lake with a complex lake shape (like that discussed in Chapter 11) we could expect considerably less focussing. Nevertheless, a form of wavetrains was observed for the round lake case shown in Fig. 5.4.

CHAPTER 6 Rotating shallow water dynamics

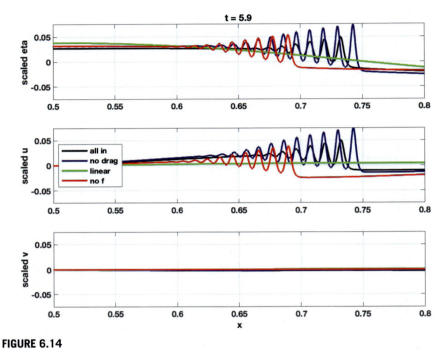

FIGURE 6.14

The detail of the wavetrain development at dimensionless time $t = 5.9$. Black – full model, blue – model without drag, green – linearized model, red – no rotation ($f = 0$). Upper panel – free surface scaled by undisturbed depth, Lower – panel u scaled by c_0.

Essentially all lake scale simulations involve some form of eddy viscosity, or some way to represent diffusion by unresolved motions. Since the models discussed so far have little inherent numerical dissipation, we can examine the issue of what we may lose if we set our eddy viscosity to be too high in detail. To do this we reconsider the seiche case from the previous examples, but now contrast the case with the bottom drag and several cases with eddy viscosity. The x-momentum equation without bottom drag and with eddy viscosity reads

$$u_t + uu_x = -g\eta_x + fv + \frac{H^2}{6}u_{xxt} + \nu_{eddy}u_{xx}.$$

ν_{eddy} is a "modeled" parameter, meaning that there is no first principles theory that specifies it. We present three values, $\nu_{eddy} = (0.1, 1, 10)$, which we label as small, medium, and large. The values do fall into the range used in some lake modeling studies, but the complexity and inherent numerical dissipation of full lake scale models, makes a quantitative comparison impossible. The relevant code including eddy viscosity is given in

```
swnh1d_weddyvisc.m
```

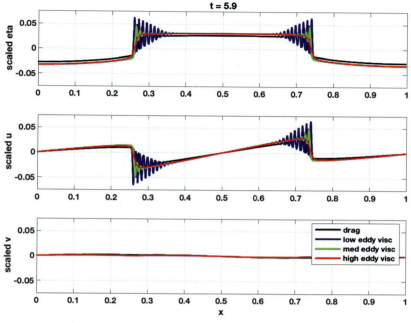

FIGURE 6.15

The seiche development problem for a dimensionless time ($t = 5.9$) at which wavetrains have developed. Black – full model, blue – $\nu_{eddy} = 0.1$, green – $\nu_{eddy} = 1$, red – $\nu_{eddy} = 10$. Upper panel – free surface scaled by undisturbed depth, Lowers – panel u and v scaled by c_0.

Figs. 6.15 and 6.16 show the seiche at dimensionless time $t = 5.9$ (the latter zooming in on the wavetrain in the range $0.65 < x < 0.775$). At this time the wavetrains have developed in most of the cases, though even on the scale of the whole domain, Fig. 6.15, differences are evident. Focusing on wavetrains we can note that changing the value of eddy viscosity qualitatively changes what is observed. The low value of viscosity (blue) has a clear wavetrain, with an amplitude that is actually even larger than the case with the bottom drag (black). The medium value of eddy viscosity (green) leads to a weaker wavetrain, with a much shorter horizontal extent. The high value of eddy viscosity (red) leads to no wavetrains at all.

While it is true that most lake scale models do not even attempt to include nonhydrostatic effects (though to be fair an increasing number of models have some nonhydrostatic capability), we have just shown that even if they did, users accustomed to high values of eddy viscosity may not observe any nonhydrostatic effects in their simulations. This situation is further muddled by the inherent numerical dissipation (essentially an invisible eddy viscosity) of many numerical schemes.

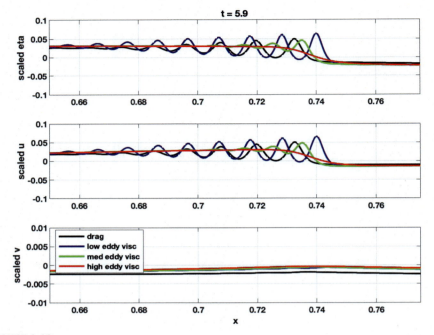

FIGURE 6.16

Detail of the wavetrains at dimensionless time ($t = 5.9$). Black – full model, blue – $\nu_{eddy} = 0.1$, green – $\nu_{eddy} = 1$, red – $\nu_{eddy} = 10$. Upper panel – free surface scaled by undisturbed depth, Lowers – panel u and v scaled by c_0.

6.5 Coding ideas

We have had considerable success with the above equations. This was due to a combination of good modeling choices (i.e. including dispersion) and good quality numerics.

In this section we discuss aspects of the numerical techniques we use. In order to keep the discussion simple we will drop the bottom stress terms, as well as rotation. Both can be included in the numerical method in a very simple way and the interested reader can see the details in the codes provided with this book.

We want to consider the equations (where subscripts are used to denote derivatives)

$$u_t = -uu_x - g\eta_x + \frac{H^2}{6}u_{txx} \tag{6.13}$$

$$\eta_t = -[(H + \eta)u]_x. \tag{6.14}$$

We will assume the bottom is flat (for convenience only) and make use of the Discrete Fourier Transform as implemented in the FFT. Mathematically the Fourier transform is defined on the whole real line as

$$\mathcal{F}(u(x)) = \bar{u}(k) = \int_{-\infty}^{\infty} u(x)\exp(-ikx)dx$$

and the restriction to a domain $-L \leq x \leq x$ and functions that are periodic on this domain allows us to use the FFT. The mathematical details can be found in a number of excellent applied mathematics texts ([6,7]), though very few are directly relevant to the present, science first setting.

The one mathematical property of the Fourier transform that we do make use of is that it converts derivatives to a multiplication,

$$\mathcal{F}(u_x) = ik\mathcal{F}(u) = ik\bar{u}(k).$$

We can take the Fourier transform of the equations above, and rearrange the momentum equation so that all the time derivative terms are on the left hand side to find

$$\left(1 + k^2\frac{H^2}{6}\right)\bar{u}_t = -\overline{uu_x} - gik\bar{\eta} \tag{6.15}$$

$$\bar{\eta}_t = -Hik\bar{u} + ik\overline{\eta u}. \tag{6.16}$$

Recall that the overbar denotes a Fourier transformed variable. Next, we follow standard notation in numerical methods and write

$$u^{(n)} = u(x, n\Delta t).$$

This allows us to use standard finite differencing to approximate the time derivative. For example the forward Euler time stepping would write

$$u_t = \frac{u^{(n+1)} - u^{(n)}}{\Delta t} + O(\Delta t).$$

This is not a great choice, but a centered second order version, often called the leapfrog method works much better

$$\frac{u^{(n+1)} - u^{(n-1)}}{\Delta t} + O(\Delta t^2).$$

The leap frog method has a bit of a bad reputation in numerical analysis circles, owing to its strange instabilities when used in purely finite difference schemes. Here, however, we are using the FFT to compute spatial derivatives and the technique is very effective. We do one mathematically clever manipulation by noticing that the chain rule of calculus tells us that

$$(u^2/2)_x = uu_x$$

112   CHAPTER 6   Rotating shallow water dynamics

so that

$$\overline{uu_x} = \frac{ik}{2}\overline{u^2}.$$

If we also write the factor due to dispersion as

$$M(k) = \frac{1}{1 + k^2 \frac{H^2}{6}} \tag{6.17}$$

the discretization after Fourier transforming can be written as a simple recipe,

$$u^{(n+1)} = u^{(n-1)} - M(k)\left[\Delta t i k \overline{u^2}^{(n)} + 2\Delta t g i k \bar{\eta}^{(n)}\right] \tag{6.18}$$

$$\bar{\eta}^{(n+1)} = \bar{\eta}^{(n-1)} - Hik\bar{u}^{(n)} + ik\overline{\eta u}^{(n)}. \tag{6.19}$$

In practice after advancing, the $\bar{u}^{(n+1)}$ and $\bar{\eta}^{(n+1)}$ variables are transformed back to physical space using the inverse FFT. It is somewhat amazing that the dispersive correction is so simple to implement (a simple multiplicative a factor that does not change with time). This is partly due to the simplifying assumption of a flat bottom (i.e. Fourier methods are superb at constant coefficient problems), and partly due to the properties of the Fourier transform (that carry over to the FFT). We will see in Chapters 9–11 that other numerical methodologies require more sophisticated methods to implement the dispersive terms.

The perceptive reader will have noticed that the technique requires the u and η fields at the present time step and one time step in the past. This is problematic for the first time step, and usually a single explicit Euler step is taken so that one has the initial condition and the first time step, and can then apply the leapfrog method from that point on.

For readers who prefer a pseudocode listing we provide this in Algorithm 1 below.

Algorithm 1 Theoretical time stepping the dispersive SW equations.

Require: parameters and initial conditions

    COMPUTE the FFT of the ICs for the RHS of the PDEs

    CALCULATE one Euler forward step to get $\bar{u}^{(1)}$ and $\bar{\eta}^{(1)}$

    COMPUTE the inverse FFT to get $u^{(1)}$ and $\eta^{(1)}$

    DEFINE $u_{now} \leftarrow u^{(1)}$, $\eta_{now} \leftarrow \eta^{(1)}$, $u_{past} \leftarrow u^{(0)}$, and $\eta_{past} \leftarrow \eta^{(1)}$

    $n \leftarrow 1$

    while n is less than number of time steps **do**

        COMPUTE the FFT of u_{now}, u_{past}, η_{now}, η_{past} for the RHS of the PDEs

        CALCULATE one leapfrog forward step to get $\bar{u}^{(n+1)}$ and $\bar{\eta}^{(n+1)}$

        COMPUTE the inverse FFT to get $u^{(n+1)}$ and $\eta^{(n+1)}$

        UPDATE $u_{past} \leftarrow u_{now}$, $\eta_{past} \leftarrow \eta_{now}$, $u_{now} \leftarrow u^{(n+1)}$, and $\eta_{now} \leftarrow \eta^{(n+1)}$

        $n \leftarrow n + 1$

    end while

In practice one typically wants to either store data or produce graphics at intermediate points and the algorithm is modified as in Algorithm 2.

Algorithm 2 Practical time stepping of the dispersive SW equations.

Require: parameters and initial conditions including number of outputs and inner steps

  COMPUTE the FFT of the ICs for the RHS of the PDEs

  CALCULATE one Euler forward step to get $\bar{u}^{(1)}$ and $\bar{\eta}^{(1)}$

  COMPUTE the inverse FFT to get $u^{(1)}$ and $\eta^{(1)}$

  DEFINE $u_{now} \leftarrow u^{(1)}$, $\eta_{now} \leftarrow \eta^{(1)}$, $u_{past} \leftarrow u^{(0)}$, and $\eta_{past} \leftarrow \eta^{(1)}$

  $n \leftarrow 1$

  while i is less than number of outputs **do**

    while j is less than number of inner time steps **do**

      COMPUTE the FFT of unow, upast, etanow, etapast for the RHS of the PDEs

      CALCULATE one leapfrog forward step to get $\bar{u}^{(n+1)}$ and $\bar{\eta}^{(n+1)}$

      COMPUTE the inverse FFT to get $u^{(n+1)}$ and $\eta^{(n+1)}$

      UPDATE $u_{past} \leftarrow u_{now}$, $\eta_{past} \leftarrow \eta_{now}$, $u_{now} \leftarrow u^{(n+1)}$, and $\eta_{now} \leftarrow \eta^{(n+1)}$

      $n \leftarrow n + 1, j \leftarrow j + 1$

    end while

    Store data or create graphical output

    $i \leftarrow i + 1, j \leftarrow 1$

  end while

The primary task of this section is to demonstrate to the reader that the same smart choices that make the mathematical modeling process successful can be applied to numerical methods. The code based on the above algorithms will not solve ALL problems, indeed it has some very clear limitations, but it will solve a subset of problems extremely efficiently. This efficiency then allows the user to concentrate on posing smart thought experiment using the code, as we did in the previous sections.

The code is also quite memory efficient, to the point where the implementation of the algorithms above evolved all the fields for the numerical experiments presented in the above sections at the same time. This is not necessary, but it does cut down on issues associated with data management. If we were presenting large two or even three dimensional simulations a different approach would likely be adopted, but for many "what if" numerical experiments, writing code that does your entire experiment leads to a far more efficient scientific computing process. Put another way, algorithms matter, but usability and user experience matter just as much.

6.6 Mini-projects

This chapter is somewhat exceptional in the sense that rather than pursuing problems beyond what is described, a reader who really wants to understand the material should carefully dissect the codes provided, and run them with their own changes to see how the results are modified.

We suggest starting with the small scale adjustment case, followed by the seiche case (with or without eddy viscosity). This is because the seiche case has a somewhat smaller number of points than the large scale adjustment case, and early explorations tend to go better when there is not a long wait time for results.

Examples of issues worth paying attention to are:

1. How are the nonlinear terms handled?
2. What is done to tame aliasing?
3. How are graphics handled?

Examples of things a beginning reader may change are to move the initial conditions, or change their functional form. A mid-level reader may want to compare quadratic and linear drag, or look at the spectra of the resulting fields. An expert reader may wish to change the method used for time-stepping or to optimize the number of FFTs used at each time step.

References

[1] P. Brandt, et al., Internal waves in the Strait of Messina studied by a numerical model and synthetic aperture radar images from ERS1/2 satellites, Journal of Physical Oceanography 27 (1997) 648–663.

[2] A. de la Fuente, et al., The evolution of internal waves in a rotating, stratified, circular basin and the influence of weakly nonlinear and nonhydrostatic accelerations, Limnology and Oceanography 53 (6) (2008) 2738–2748.

[3] P.K. Kundu, I.M. Cohen, Fluid Mechanics, 4th ed., Elsevier Academic Press, 2008.

[4] Clotilde Le Quiniou, et al., Copepod swimming activity and turbulence intensity: study in the Agiturb turbulence generator system, The European Physical Journal Plus 137 (2) (2022) 1–14.

[5] R.J. Leveque, Finite Volume Methods for Hyperbolic Problems, Cambridge University Press, 2002.

[6] Brad G. Osgood, Lectures on the Fourier Transform and Its Applications, vol. 33, American Mathematical Soc., 2019.

[7] L.N. Trefethen, Spectral Methods in MATLAB, Society for Industrial and Applied Mathematics, 2000.

[8] G.B. Whitham, Linear and Nonlinear Waves, Wiley-Interscience, 1999.

CHAPTER

7

Understanding complex dynamics in two and three dimensions

CONTENTS

7.1 Turbulence closures: the idea.. 115
7.2 Turbulence closures: Navier Stokes... 118
7.3 Large-eddy simulations (or LES) 120
7.4 Slow-fast systems 124
7.5 Mini-projects 129
References.. 131

In this chapter we try to provide an introduction to a more systematic approach to simulations that involve more than one dimension. In practice, of course, these are precisely the simulations that tell us information about a particular geographical location. At the same time they are potentially the most complicated, and also possibly the most distant from theory.

We have used eddy viscosity as a blunt instrument in the previous chapter. Here we develop the ideas behind turbulent closures and eddy viscosity. This is done using a one dimensional model equation called the Burgers equation. We then move onto a discussion of the algebra of turbulent closures in the more complex context of the Navier-Stokes equations. We return to the simpler context of the Burgers equation to provide a more modern take on turbulence simulation called large eddy simulation or LES. This idea involves a spatially averaged version of the variables of fluid mechanics. The algebra is similar to, albeit a touch more complicated than Reynolds averaging. For this reason the idea is demonstrated on the Burgers equation and then simply quoted for the Navier Stokes equations. This allows us to focus on the main task needed for practical applications, which is setting the eddy viscosity. Finally we switch gears from turbulence theory and return to the shallow water equations on the f-plane and discuss the notion of fast-slow systems, with some examples from simulations.

7.1 Turbulence closures: the idea

It is an observational fact that most fluid motions that occur in the natural world are highly irregular. Interestingly, how precisely fluid motions become irregular has re-

Physics and Ecology in Fluids. https://doi.org/10.1016/B978-0-32-391244-0.00017-6
Copyright © 2023 Elsevier Inc. All rights reserved.

115

116 CHAPTER 7 Understanding complex dynamics

ceived more historical attention than the irregular state itself. For now let's assume that "irregular", "chaotic", "random" and similar adjectives apply to the verbal description of the flow. Mathematically, it is thus a far more reasonable idea to seek a description of the average state of the fluid as opposed to a complete time dependent description. This point of view goes back to Osborne Reynolds in the 19th century, and before the advent of cheap, high speed computation was pretty much the only way to look at complicated flows. We will see that there is a catch. The reader should note that this notion of mean and fluctuation makes no assumption about the smallness of the fluctuations.

To begin a first-principles quantitative discussion we would require a proper mathematical definition of random variables. This is well beyond the scope of the present book, so we will proceed in an *ad hoc* manner leaving the interested readers to pursue the matter at depth on their own. There are actually two types of averages, time averages defined as

$$\bar{f} = \lim_{T \to \infty} \frac{1}{T} \int_{t_0}^{t_0+T} f(s)\,ds \tag{7.1}$$

and the so-called ensemble average over experimental realizations

$$\langle f \rangle = \lim_{N \to \infty} \frac{1}{N} \sum_{n=1}^{\infty} f_n(t). \tag{7.2}$$

In formal statistical mechanics the equivalence, or lack thereof, between the two averages is the concern of ergodic theorems. For this section consider the ensemble average only, and note that in experiments, N might be several realizations as opposed to an actual limit to infinity.

To keep the algebra as simple as possible, we turn to the Burgers equation ([3]); a model scalar PDE for the total variable $u^T(x, t)$,

$$u_t^T + u^T u_x^T = v u_{xx}^T. \tag{7.3}$$

The left hand side of the equation is the material derivative in one-dimension, and hence this system has some analogy with the Navier-Stokes equations (though is obviously much simpler). We will consider the full Navier-Stokes equations in the next section. We make an additive decomposition between the mean and fluctuating parts,

$$u^T = U + u, \tag{7.4}$$

where U is some slowly varying mean velocity and u is the rapidly fluctuating part of the velocity so that $\langle u \rangle = 0$. Note that $\langle u^T \rangle = U$, for obvious reasons. If we follow Reynolds and use the decomposition (7.4) in (7.3) and average we find first that

$$U_t + \langle u_t \rangle + U U_x + U \langle u_x \rangle + \langle u \rangle U_x + \langle u u_x \rangle = v U_{xx} + v \langle u_{xx} \rangle.$$

7.1 Turbulence closures: the idea **117**

However $\langle u \rangle = 0$ and it is generally assumed that $\langle . \rangle$ commutes with partial derivatives (this is obvious if we have a large, but finite number of ensembles N). This simplifies the equation a great deal so that

$$U_t + U U_x = \nu U_{xx} - \frac{1}{2} \langle uu \rangle_x. \tag{7.5}$$

For the right hand side we have used the product rule to write

$$\langle u^2 \rangle_x = 2 \langle uu_x \rangle.$$

Notice that even though $\langle u \rangle = 0$, we have no guarantee that $\langle uu \rangle = 0$, and indeed the term $\langle uu \rangle_x$ is generally not zero and acts as a source/sink for the mean flow.

Thus we were able to derive an equation for the average velocity that did not involve the fluctuating parts **except** in the term $\langle uu \rangle_x$. Sadly, we do not have an equation for this term and hence the system is not closed. What if we were to try to derive one? Multiply (7.3) by u^T, use the Reynolds decomposition (7.4), and average. The algebra is a bit nasty, but if we use the fact that $\langle u^T \rangle = U$ and $\langle u \rangle = 0$ the result can be simplified to read

$$\frac{1}{2} \langle u^2 \rangle_t = -\frac{1}{2} (U^2)_t - U_x \langle u^2 \rangle + U \langle u^2 \rangle_x - U^2 U_x + \nu U U_x + \nu \langle uu_{xx} \rangle - \langle u^2 u_x \rangle. \tag{7.6}$$

On the left hand side we have the time derivative of $\langle uu \rangle$ and all the terms but the last are combinations of U, $\langle u^2 \rangle$ and their various derivatives. The last term however involves what is called a triple correlation, which again, is not zero, and we have no equation for! Were we to multiply (7.3) by u_T^2, then use the Reynolds decomposition (7.4) and average, we would find that this new equation for $\langle u^3 \rangle$ would involve another higher product $\langle u^4 \rangle$, and hence we would be no closer to closing the system! This is in fact a general result, and goes by the name of the **Turbulence Closure Problem**.

We could however **model** the effects of the turbulence. It is experimentally observed that one of the main bulk effects of turbulence is a massive increase in diffusivity. Indeed without turbulence in the air, a pan of tasty cookies that just came out of the oven would not tempt you while you were out of sight, studying in your room, for more than a day. This is because molecular diffusion acts incredibly slowly when compared to the effective diffusion of turbulent motions. Thus we could define a phenomenological constant of "eddy viscosity" $\nu_T \gg \nu$ and write

$$-\frac{1}{2} \langle u^2 \rangle_x = \nu_T U_{xx} \tag{7.7}$$

so that the phenomenological closed system reads

$$U_t + U U_x = \nu_T U_{xx}. \tag{7.8}$$

118 CHAPTER 7 Understanding complex dynamics

The fact that the eddy viscosity is constant is difficult to justify, except perhaps as an appeal to Ockham's razor. However, as we saw in the previous chapter, a constant eddy viscosity is very easy to incorporate into standard numerical schemes.

7.2 Turbulence closures: Navier Stokes

With the core idea of the previous section in hand we consider the three dimensional Navier Stokes and present the algebra of the turbulence closure. We begin with the constant density Navier Stokes equations

$$\frac{D\vec{u}}{Dt} = -\frac{1}{\rho_0}\nabla p + \nu\nabla^2\vec{u}$$

and the Conservation of Mass (which is really the Conservation of Volume)

$$\nabla \cdot \vec{u} = 0.$$

Before turning to any turbulence theory we make use of the Conservation of Mass to rewrite the $\vec{u} \cdot \nabla\vec{u}$ term in the Navier Stokes equations. The trick is best show with a particular example, so consider the equation for the u component of $\vec{u}$.

$$\vec{u} \cdot \nabla u = uu_x + vu_y + wu_z$$

where we have used subscripts for partial derivatives on the right hand side. The Conservation of Mass in the same notation reads

$$u_x + v_y + w_z = 0$$

so that

$$(uu)_x + (vu)_y + (wu)_z = u(u_x + v_y + w_z) + uu_x + vu_y + wu_z$$

and hence

$$\vec{u} \cdot \nabla u = \nabla \cdot (u\vec{u}). \tag{7.9}$$

This means we can rewrite the Navier-Stokes equations in the alternative form

$$\frac{\partial\vec{u}}{\partial t} + \nabla \cdot \mathbf{R} = -\frac{1}{\rho_0}\nabla p + \nu\nabla^2\vec{u} \tag{7.10}$$

where

$$\mathbf{R} = \begin{pmatrix} uu & vu & wu \\ uv & vv & wv \\ uw & vw & ww \end{pmatrix}. \tag{7.11}$$

7.2 Turbulence closures: Navier Stokes 119

With this result in hand we make the Reynolds decomposition

$$\vec{u} = \vec{U} + \vec{u}', \tag{7.12}$$

with a similar definition for p, and we note that derivatives and the averaging process commute (i.e. averaging then differentiating is the same as differentiating then averaging). We wish to write down a set of equations for the averaged quantities and, based on our work in the last section we expect to be stymied by pairs of fluctuating quantities being averaged.

The source of any problems will be the $\mathbf{R}$ matrix and so let's consider that one in detail. Notice that $\langle Uv \rangle = 0$ because U is an average quantity and hence passes outside the angled brackets, and $\langle v \rangle = 0$. In fact this is the same for any product of an average and fluctuating quantity. This greatly simplifies the averaged $\mathbf{R}$ so that

$$\langle \mathbf{R} \rangle = \begin{pmatrix} UU + \langle uu \rangle & VU + \langle vu \rangle & WU + \langle wu \rangle \\ UV + \langle uv \rangle & VV + \langle vv \rangle & WV + \langle wv \rangle \\ UW + \langle uw \rangle & VW + \langle vw \rangle & WW + \langle ww \rangle \end{pmatrix}.$$

The averaged Navier Stokes equations can thus be written as

$$\frac{\partial \vec{U}}{\partial t} + \vec{U} \cdot \nabla \vec{U} = -\frac{1}{\rho_0} \nabla P + \nu \nabla^2 \vec{U} - \nabla \cdot \begin{pmatrix} \langle uu \rangle & \langle vu \rangle & \langle wu \rangle \\ \langle uv \rangle & \langle vv \rangle & \langle wv \rangle \\ \langle uw \rangle & \langle vw \rangle & \langle ww \rangle \end{pmatrix}.$$

With only a minor rewrite these can be put into a very useful, and suggestive form. If we label the viscous stress tensor based on the average flow as

$$\tau_v = \mu(\nabla \vec{U} + [\nabla \vec{U}]^T)$$

we get the system

$$\rho_0 \frac{\partial \vec{U}}{\partial t} + \rho_0 \vec{U} \cdot \nabla \vec{U} = -\nabla P + \nabla \cdot \tau_v + \nabla \cdot (-\rho_0 \mathbf{R}). \tag{7.13}$$

On the left hand side we have the rate of change of momentum with time and the advection of momentum by the average current. On the right hand side we have the force due to the gradient of average pressure and two further terms. One is the viscous stress based on the averaged flow and the other is new. It specifies the transport of fluctuation momentum by the fluctuations $(-\rho_0 \mathbf{R})$. This is the piece that yields the turbulence closure problem, meaning we cannot write down a description of the averaged state in terms of the average alone. The new term goes by the term **Reynolds stress** and in many situations in the natural world is dominant.

To close the problem the eddy viscosity is introduced so that

$$\rho_0 \mathbf{R} = \mu_E(\nabla \vec{U} + [\nabla \vec{U}]^T) \tag{7.14}$$

120 **CHAPTER 7** Understanding complex dynamics

where μ_E must be determined in some *ad hoc* way. Models that do this are often put into a catch all category of Reynolds Averaged Navier Stokes models, or **RANS**.

There is a bit of a contradictory nature to how these models are used in practice since one typically runs a single instance of dynamics with eddy viscosity (often quite a large value) but the description is for the average. Nevertheless, RANS continues to be the dominant form of simulation for natural waters.

7.3 Large-eddy simulations (or LES)

There is an alternative point of view to Reynolds averaging. It begins with the idea that a given simulation can only resolve some scales, and hence the fluid motion should be divided into resolved and sub-grid portions. Mathematically, this can be accomplished by filtering the variables.

In this section we examine this point of view using the Burgers equation to make the algebra as simple as possible. A far more complete picture is present in [4].

Begin by defining the filtered variable

$$\langle u \rangle^{LES}(x,t) = \int_D F(y-x)u(y,t)dy \tag{7.15}$$

where the subscript LES is added to distinguish this averaging process from the Reynolds averaging of the previous section. Also, in the above y is a dummy variable of integration and F is called the averaging kernel. For concreteness the reader can think of a Gaussian form of F and the argument $y - x$ shifts the Gaussian to be centered in the right place. Two examples of averaging a function are given in Figs. 7.1 and 7.2 for a narrow and wider filtering kernel function. The original function is

$$F(x) = \sin\left(\frac{2\pi x}{10}\right) + \sin\left(\frac{2\pi x}{1} + \phi_1\right) + \sin\left(\frac{2\pi x}{2} + \phi_2\right) + \sin\left(\frac{2\pi x}{0.7} + \phi_3\right) \tag{7.16}$$

where ϕ_i are random phase shift that ensure there is no systematic pile up of the function value near $x = \pi/2$ and $x = 3\pi/2$. The code used to generate the figures is

```
les_basics.m
```

The narrow filtering kernel function effectively only "sees" a more localized area and hence, while it filters the faster oscillating portions of the function, two clear wavelengths are visible in Fig. 7.1. The wider window is more effective at smearing the finer features and for this reason Fig. 7.2 shows a filtered signal that is essentially only the sinusoid of period 10.

There are many other filters that can be chosen. If one were to want to be particularly harsh one could choose the so-called box car filter

$$F(x;\delta) = \frac{1}{\delta}[H(x+0.5\delta) - H(x-0.5\delta)],$$

7.3 Large-eddy simulations (or LES)

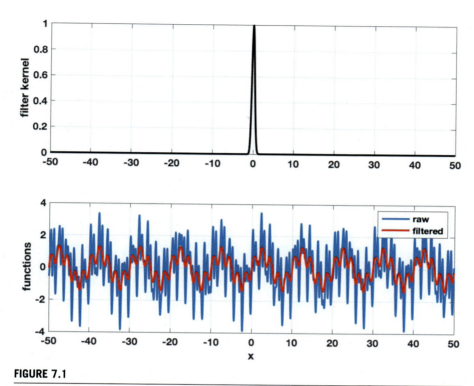

FIGURE 7.1

Narrow filtering function example. Upper panel – filtering kernel function centered at zero, lower panel – raw (blue) and filtered function (red).

where $H(.)$ is the Heaviside step function, so that the filter sets the field to zero outside of the interval $[x - 0.5\delta, x + 0.5\delta)]$. Because so much of turbulence theory progress has involved periodic domains, many formulations of LES are carried out in Fourier space, with the filter begin a function of k. The Gaussian is a special filter, since in both physical space and Fourier space it takes same functional form.

To use LES in practice we must filter the governing equation. The primary difference is that while repeated Reynolds averaging does not do anything new after the first average is taken, repeatedly filtering yields new functions. Thus for example

$$\langle\langle u \rangle\rangle = \langle u \rangle$$

but

$$\langle\langle u \rangle^{LES}\rangle^{LES} \neq \langle u \rangle^{LES}.$$

Consider the Burgers equation

$$u_t + u u_x = \nu u_{xx}$$

122 CHAPTER 7 Understanding complex dynamics

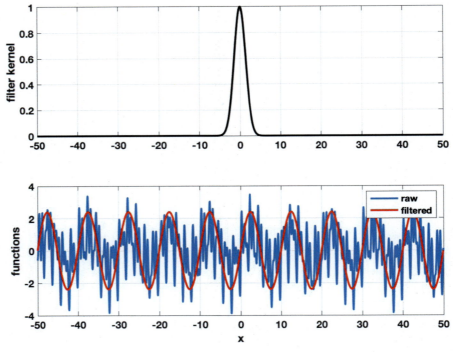

FIGURE 7.2

Wider filtering function example. Upper panel – filtering kernel function centered at zero, lower panel – raw (blue) and filtered function (red).

notice that $(u^2/2)_x = u u_x$ and take the average of both sides. This gives

$$\langle u \rangle_t^{LES} + \frac{1}{2} \langle u^2 \rangle_x^{LES} = v \langle u \rangle^{LES}.$$

The question now is how to treat the average of the quadratic term. Let us write u as the averaged part plus a fluctuation part defined as

$$u' = u - \langle u \rangle^{LES}.$$

Using this definition we write

$$\langle u^2 \rangle_x^{LES} = \langle \langle u \rangle^{LES} \langle u \rangle^{LES} \rangle_x^{LES} + 2 \langle \langle u \rangle^{LES} u' \rangle^{LES} + \langle u' u' \rangle^{LES}. \qquad (7.17)$$

This is more complex than the Reynolds stress of the previous section, but can still be interpreted. The first term on the right hand side is the effect of the averaged state, the second is the interaction between the averaged state and the fluctuations, and the third is the analogy of the Reynolds stress and represents the effects of the fluctuations. In practice both the second and third terms are replaced by an eddy viscosity-like term.

7.3 Large-eddy simulations (or LES)

Large eddy simulation is, in many cases, not practically different from RANS, but it is conceptually different. RANS offers means for building eddy viscosities that are not constant, e.g. the well known $k - \epsilon$ model used in engineering, as discussed in [2]. However, LES allows for the opportunity to develop models that explicitly account for the grid size. In particular, when resolution is increased, the eddy viscosity can be decreased. This is because more scales are explicitly resolved, and therefore do not need to be damped out.

Here we present the averaged Navier-Stokes equations, leaving the details of the procedure for the reader.

$$\langle \vec{u} \rangle^{LES} + \nabla \cdot \langle \mathbf{R} \rangle^{LES} = -\frac{1}{\rho_0} \nabla \langle p \rangle^{LES} + \nu \nabla^2 \langle \vec{u} \rangle^{LES} + \nabla \cdot \tau^{\mathbf{LES}} \tag{7.18}$$

Here

$$\langle \mathbf{R} \rangle^{LES} = \begin{pmatrix} \langle u \rangle^{LES} \langle u \rangle^{LES} & \langle v \rangle^{LES} \langle u \rangle^{LES} & \langle w \rangle^{LES} \langle u \rangle^{LES} \\ \langle u \rangle^{LES} \langle v \rangle^{LES} & \langle v \rangle^{LES} \langle v \rangle^{LES} & \langle w \rangle^{LES} \langle v \rangle^{LES} \\ \langle u \rangle^{LES} \langle w \rangle^{LES} & \langle v \rangle^{LES} \langle w \rangle^{LES} & \langle w \rangle^{LES} \langle w \rangle^{LES} \end{pmatrix}$$

and $\tau^{\mathbf{LES}}$ represents all the contributions from the advective terms that CANNOT be represented in terms of the filtered flow only. It is often written, somewhat confusingly as

$$\tau^{\mathbf{LES}} = -\begin{pmatrix} \langle uu \rangle^{LES} & \langle vu \rangle^{LES} & \langle wu \rangle^{LES} \\ \langle uv \rangle^{LES} & \langle vv \rangle^{LES} & \langle wv \rangle^{LES} \\ \langle uw \rangle^{LES} & \langle vw \rangle^{LES} & \langle ww \rangle^{LES} \end{pmatrix}$$

$$+ \begin{pmatrix} \langle u \rangle^{LES} \langle u \rangle^{LES} & \langle v \rangle^{LES} \langle u \rangle^{LES} & \langle w \rangle^{LES} \langle u \rangle^{LES} \\ \langle u \rangle^{LES} \langle v \rangle^{LES} & \langle v \rangle^{LES} \langle v \rangle^{LES} & \langle w \rangle^{LES} \langle v \rangle^{LES} \\ \langle u \rangle^{LES} \langle w \rangle^{LES} & \langle v \rangle^{LES} \langle w \rangle^{LES} & \langle w \rangle^{LES} \langle w \rangle^{LES} \end{pmatrix} .$$

The reason for the confusion is that unlike in our work for the Burgers equation, it isn't clear what part is purely due to the unresolved scales and what part is the interaction between the filtered (or resolved) scales and the unresolved scales.

Nevertheless, if one wishes to make a computational model, one needs to choose an *ad hoc* model for $\tau^{\mathbf{LES}}$. One such model is called the **Smagorinsky model**. We write the eddy stress as

$$\tau^{\mathbf{LES}} = \nu^{LES} \left(\nabla \langle u \rangle^{LES} + [\nabla \langle u \rangle^{LES}]^T \right)$$

where the term in the brackets on the right hand side is two times the rate of strain tensor $\langle \mathbf{S} \rangle^{LES}$ based on the filtered (or resolved) flow. If the grid spacing is denoted by Δ then the eddy viscosity is built from the filtered rate of strain according to the

124 CHAPTER 7 Understanding complex dynamics

formula

$$v^{LES} = C\Delta^2 \sqrt{2 \sum_{i,j=1}^{3} \langle S_{ij} \rangle^{LES} \langle S_{ij} \rangle^{LES}} \tag{7.19}$$

where C is a constant. This says that the Smagorinsky eddy viscosity is large (small) when the rate of strain of the resolved field is large (small). In practice, this is not a perfect model (more complex models can be found in [2]). However, the model presented above is the simplest model for an eddy viscosity that keeps the extra viscosity due to turbulence small in some regions and does so in a rational way.

7.4 Slow-fast systems

The intuition on turbulence modeling developed in the two previous sections is built on the basis of the three dimensional Navier-Stokes equations. In essence, it all comes down to the nonlinear terms, $\vec{u} \cdot \nabla \vec{u}$, and their role during whichever averaging procedure is used.

When modeling the flow in natural waters one vital consideration is the discrepancy of scales between the large horizontal extent of the basin and the small depth. There is no reason to believe that what is essential for small scale turbulence is essential on the basin scale. In this section we take a purely numerical point of view on this question. We perform idealized simulations which we then examine using the ideas from the previous section on large eddy simulations.

Recall that in Chapter 5 (5.6) we discussed the possible wave forms in the shallow water equations on the f-plane: for example the dispersion for rotation modified gravity waves (5.6), and the decaying away from the coast form of the Kelvin waves (5.7). When boundaries are neglected, we found rotation-modified gravity waves, the so-called Poincaré waves, and a mode for which the frequency was zero. This corresponds to the geostrophic balance state, in which flow is along lines of constant pressure (or equivalently along the lines of constant free surface height η). You may recall that this picture was obtained by linearizing, and hence we could ask what happens in a simulation of the full system.

We explore this by considering a state that is a combination of a geostrophic balance and a short wave perturbation. There are many possible choices, but we choose a set up that will be similar to the ad hoc velocity fields of Chapter 3 (in particular the 2D cases of Section 3.5). The basic idea is to write

$$\eta(x, y, t = 0) = \eta_G(x, y) + \eta_{UB}(x, y)$$

so that

$$(u, v)(x, y, t = 0) = \frac{g}{f}(-\eta_y, \eta_x)$$

where subscripts on the right hand side denote partial derivatives. This means the initial velocities balance the initial η_G field, but the η_{UB} field is unbalanced (hence the subscript UB). Since it is generally accepted that the balanced state is associated with large scales, while the unbalanced state is associated with short scales we take

$$\eta_G = \frac{H}{30} \sin\left(\frac{\pi x}{L_x}\right) \sin\left(\frac{\pi y}{L_y}\right) \tag{7.20}$$

while

$$\eta_{UB} = \frac{H}{10} \sin\left(\frac{8\pi x}{L_x}\right) \sin\left(\frac{8\pi y}{L_y}\right) \tag{7.21}$$

where the model lake is rectangular with an extent of $[-L_x, L_x] \times [-L_y, L_y]$. See Table 7.1 for a full list of simulation parameters.

Table 7.1 Parameters for the Slow-Fast simulation.

Label	(N_x, N_y)	(L_x, L_y) (km)	H (m)	g (m s^{-1})	f (s^{-1})
Slow-fast	$(256, 256)$	$(5, 5)$	20	0.0981	1.45×10^{-4}

The initialization consists of four counter-rotating cells in geostrophic balance and a perturbation that is free to break down into a number of propagating wave fronts. There are a number of possibly quantities to examine, but we choose to create four panel figures with the free surface displacement, the kinetic energy, the vorticity, and the divergence. The idea behind this choice is that the η field should show us how the unbalanced portion of the flow behaves, while the kinetic energy is expected to be dominated by the geostrophic flow. The vorticity and divergence are standard variables associated with the balanced and unbalanced motions. The basic argument for this considers the linearized shallow water equations on the f-plane

$$u_t - fv = -g\eta_x$$
$$v_t + fu = -g\eta_y$$

cross differentiates

$$u_{ty} - fv_y = -g\eta_{xy}$$
$$v_{tx} + fu_x = -g\eta_{yx}$$

and subtracts, noting that the vorticity is defined as $\omega = -u_y + v_x$, to find

$$\omega_t = -f\nabla \cdot \vec{u}.$$

In the geostrophic approximation the flow is steady, so that $\omega_t = 0$, and when the flow is close to geostrophy the expectation is that ω_t is very small. This in turn means that slowly changing flows have essentially no divergence. Thus divergence is clearly

126 **CHAPTER 7** Understanding complex dynamics

associated with fast motions, while vorticity is associated with slow (or balanced) motions. There are more complete decompositions available in the literature ([1]). We stick with the intuitive choice of vorticity and divergence in the discussion below.

The 2D simulation is complex enough so that we have chosen to split the code between the main, numerical solver and a separate codes for analysis and graphics. The code used to solve the equations is

```
swnh2d_book.m
```

with the graphics given in

```
plot2d.m
```

and some of the LES analysis given in

```
plot_les_like.m
```

Fig. 7.3 shows the early evolution after about three quarters of an hour, or dimensionless time 0.406. Time is nondimensionalized by the rotation parameter f. It is clear that the variables associated with the unbalanced motions (η and $\nabla \cdot \vec{u}$) manifest completely different patterns from those associated with the balanced motion (kinetic energy and vorticity). The free surface consists of medium scale cells of varying polarity that have been twisted by the vortical motions in each of the cells. In contrast the vorticity shows the alternating pattern of the large scale cells, with small scale striations due to the unbalanced motions. Intuitively we would thus say that the two sets of motions interact, but perhaps not particularly strongly. It is also clear that the spectral method used is very efficient at retaining flow symmetry, even for the short scale motions (something we might not expect from a model with a large amount of internal numerical dissipation and dispersion).

After roughly 1.5 hours, or at dimensionless time 0.811, the flow reaches the state shown in Fig. 7.4. It can be seen that the unbalanced flow field has led to the generation of wave fronts throughout the domain (perhaps seen most clearly in the divergence field). The large counter-rotating vortices retain their form, though appear to tilt (especially evident in the two blue vortices).

The state after just over three hours, or at dimensionless time 1.62, is shown in Fig. 7.5. It can be seen that the observations made on the basis of Fig. 7.4 remain relevant, with the counter-rotating vortices slowly moving, while the unbalanced motions continue to consist of a large number of wave fronts that, at least to the naked eye, appear to be well-resolved in the simulation. On the one hand, this can be considered a success. The simulation was run on my laptop and took about 15 minutes to complete. We could thus certainly do much, much better in terms of resolution by running on a moderately-sized workstation overnight. However, we would still be in the uncomfortable situation of not having made much contact with the world of theory described in the previous two sections.

There are two possible ways to ameliorate this situation and make contact with the ideas of large-eddy simulation (LES). We could run a simulation with only the

7.4 Slow-fast systems   127

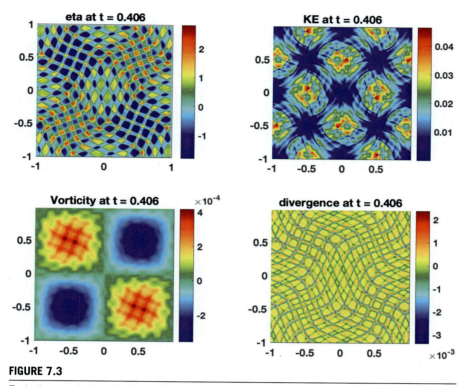

FIGURE 7.3

Early time evolution of balanced and unbalanced flow. (a) Free surface height, (b) kinetic energy, (c) vorticity, (d) divergence.

balanced motions only, or we could perform the averaging procedure on the existing data set. It turns out that this is a bit of a false dichotomy, since LES is often discussed "in Fourier space" so that averaging means retaining only the Fourier components up to some cut-off wave number. The interested reader should pursue a more complete treatment of LES, for example in [2,4].

Here we consider the simulation carried out with only the initial balanced state as the filtered flow, and the sub-grid scale flow to be the difference between our full simulation and the balanced-only or "Slow" simulation.

Fig. 7.6 shows the vorticity and divergence fields for the full simulation (upper row) and balanced-only simulation (lower row). In the panel titles we have non-dimensionalized the time by multiplying by f the rotation parameter. It is thus clear that the simulation has run long enough so that the Earth's rotation plays a clear role. The vorticity fields are quite similar with large scale patterns faithfully reproduced by the balanced-only simulation. The divergence fields are fundamentally different. Indeed for the balanced-only simulation the divergence needs to be multiplied by a factor of 100 to show anything at all! There are no wave fronts evident and the only

CHAPTER 7 Understanding complex dynamics

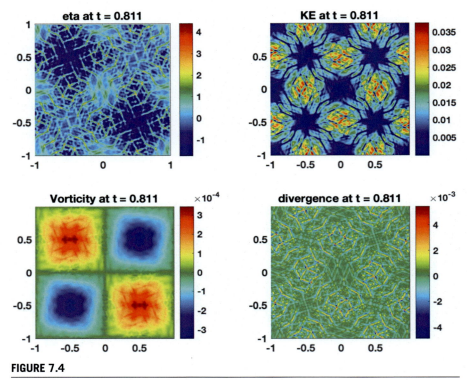

FIGURE 7.4

Medium time evolution of balanced and unbalanced flow. (a) Free surface height, (b) kinetic energy, (c) vorticity, (d) divergence.

spontaneous creation of unbalanced motions is restricted to the central regions of the vortex. This is quite reasonable since no care was taken to ensure that the form of our vortices represented those that would naturally occur.

This is a very useful result. On the one hand it suggests that were one to ONLY be interested in the slow, vortical motions, and not in the fast wave-like motions, then running coarse models should be OK. In fact, one might imagine that some form of eddy viscosity might actually be useful in making sure the flow does not misbehave at scales that approach the grid scale. However, there is a counterpoint in that any physical phenomenon thought to depend on fast motions cannot be successfully modeled by the balanced-only set up. In many natural waters, internal waves are an integral part of the inner workings of mixing and transport (especially on slopes) and our results suggest that models that properly represent such waves require both good resolution and low numerical dissipation. This issue is subtle and thorny, hence likely to keep the modelers of natural waters busy for the coming decades.

There is one final useful exercise we can carry out using our balanced-only flow: we can compute the effective eddy viscosity of the Smagorinsky model. Since we

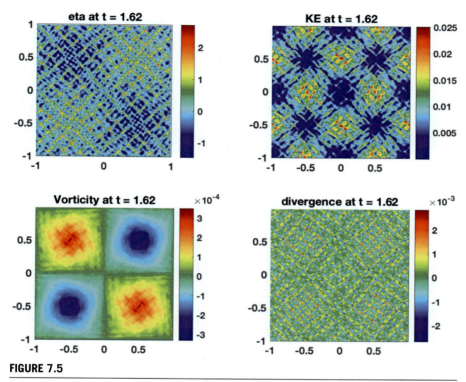

FIGURE 7.5

Late time evolution of balanced and unbalanced flow. (a) Free surface height, (b) kinetic energy, (c) vorticity, (d) divergence.

are not interested in the precise value, merely the spatial distribution, we ignore the constants in front of the square root in (7.19).

Fig. 7.7 shows the result induced by the balanced-only flow at dimensionless time $t = 1.62$. It can be seen that the diagnosed eddy viscosity is highest in locations that have little to do with the features that the eye is drawn to in Fig. 7.5 (the reader is reminded that in an LES model the small scale features are assumed to not be resolved). Indeed, for a flow like the one shown in Fig. 7.5 a far more sensible choice might be an eddy viscosity based on the divergence field. However, this field is (by definition) unavailable in a simulation that chooses to only resolve the balanced flow. Thus, in the end, there is no short cut to simulations that resolve critical features.

7.5 Mini-projects

Much like the previous chapter, a significant amount can be learned from digging into the codes provided. Because graphics are an integral part of how two dimen-

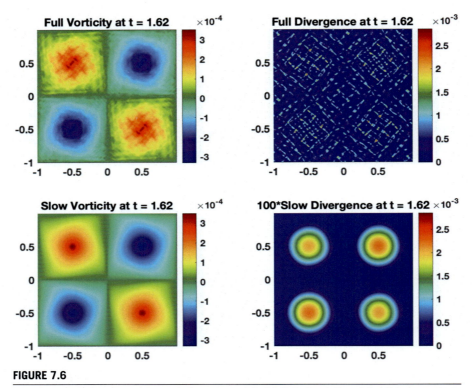

FIGURE 7.6

Late time evolution of vorticity and divergence. (a), (b) Full simulation, (c), (d) balanced-only or "Slow" simulation.

sional fields are viewed some readers may actually find it more profitable to focus on *plot2d.m* as far as their code parsing efforts are concerned.

Advanced coders may wish to decrease the number of steps taken and run the MATLAB® code profiler on *swnh2d_book.m* to see which parts of the code take the most time.

1. Use the *les_basics.m* code to show that repeatedly LES filtering does yield different functions, in contrast to repeatedly Reynolds averaging.
2. Build a code to construct an ensemble of functions, and use this code to show $\langle f \rangle$ and $\langle (f - \langle f \rangle)^2 \rangle$ for various choices of f and various numbers of ensembles.
3. Consider the shallow water equations without rotation and without the dispersive correction. Perform Reynolds averaging on these equations. You will have to use the equation of mass in order to produce Reynolds stress terms. Discuss the physics of any new terms you find. A word of warning: the algebra is difficult here.

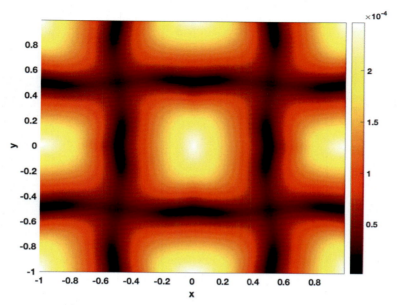

FIGURE 7.7

The spatially varying components of the Smagorinsky eddy viscosity based on the balanced-only flow at a dimensionless time $t = 1.62$.

4. Modify the initial conditions for the $swnh2d\_book.m$ code so that the difference in scale between the slow and fast motions is smaller and comment on the evolution.
5. Increase, then decrease f by a factor of 10 in $swnh2d\_book.m$ code and comment on the evolution in the two cases.
6. Using a reference of your choice (e.g. [2]) discuss the turbulent Kinetic Energy, or TKE, and the equation for its evolution.
7. Using a reference of your choice (e.g. [2]) discuss a standard RANS closure like the $k - \epsilon$ model. It makes sense to do this project after doing project 5.

References

[1] Matthew R. Ambacher, Michael L. Waite, Normal mode spectra of idealized baroclinic waves, Journal of the Atmospheric Sciences 77 (3) (2020) 813–833.
[2] Stephen B. Pope, Turbulent Flows, Cambridge University Press, 2000.
[3] G.B. Whitham, Linear and Nonlinear Waves, Wiley-Interscience, 1999.
[4] John C. Wyngaard, Turbulence in the Atmosphere, Cambridge University Press, 2010.

CHAPTER

Modeling motion in the vertical

8

CONTENTS

8.1 Stratified fluid dynamics in the x-z plane 133
8.2 The idealized internal seiche: derivation 134
8.3 The idealized internal seiche: transport... 138
8.4 The internal seiche: arbitrary stratifications... 142
8.5 The internal seiche: finite amplitude and wavetrains 147
8.6 Nonlinearity due to biological behavior 151
8.7 Mini-projects 163
References... 164

8.1 Stratified fluid dynamics in the x-z plane

The previous three chapters presented the commonly adopted "top down" view of hydrodynamics. The reason for this is simple: most natural bodies of water that have been studied have a very large lateral extent, compared to their depth. This has led to important (and sometimes opaque) mathematical short cuts. We have presented a few of these (e.g. the issue of RANS versus LES when modeling turbulent flows), in what we hope is a manner that piques the reader's interest to challenge some of the "standard" assumptions.

In this chapter we turn to another commonly adopted approximation: the side-view. It is adopted for a somewhat different reason, namely that numerical models of a natural body of water cannot resolve all the scales of motion. When vertical motions are important, simplifications are possible by neglecting the details of the basin shape, and concentrating on idealizations that can couple hydrodynamic processes (often nonhydrostatic hydrodynamic processes) and biologically relevant variables. We present a number of these in this chapter.

Throughout this chapter we will consider motion in the x-z plane. We will neglect rotation and viscosity and adopt the Boussinesq approximation. The governing stratified Euler equations read

$$\rho_0 \frac{D\vec{u}}{Dt} = -\nabla p - \rho g \hat{k} \qquad (8.1)$$

$$\nabla \cdot \vec{u} = 0 \qquad (8.2)$$

$$\frac{D\rho}{Dt} = 0 \qquad (8.3)$$

Physics and Ecology in Fluids. https://doi.org/10.1016/B978-0-32-391244-0.00018-8
Copyright © 2023 Elsevier Inc. All rights reserved.

133

134 CHAPTER 8 Modeling motion in the vertical

The essence of the Boussinesq approximation is that density changes ONLY matter in the buoyancy term. The Conservation of Mass reduces to the statement of a divergence free velocity field, and the thermodynamics simplifies so that the density is conserved following a fluid particle.

In many temperate bodies of water the density can be written in a particularly instructive way, to represent the fact that less dense (i.e. warm) water is found to overlie more dense (i.e. cold) water:

$$\rho(\vec{x}, t) = \rho_0 \left(\bar{\rho}(z) + \rho'(\vec{x}, t) \right). \tag{8.4}$$

Here ρ_0 is the reference density (typically around $1,000 \, \mathrm{kg \, m^{-3}}$), $\bar{\rho}(z)$ represents the stratification which is often taken to be fixed (or to evolve on such slow time scales that these changes can be ignored), and ρ' is taken to represent all the time dependent changes to density. The fact that the fluid is stably stratified, or with less dense fluid overlying more dense fluid, is represented by the assumption $\bar{\rho}(z) \geq 0$.

8.2 **The idealized internal seiche: derivation**

The simplest side view set up is that of a lake with a shape like a giant aquarium. The domain has a fixed length L and a fixed depth H. We assume the fluid is stably stratified and our first order of business will be to simplify the stratified Euler equations to account for our two-part density.

The density equation simplifies right away to read

$$\frac{D\rho'}{Dt} + w \frac{d\bar{\rho}}{dz} = 0.$$

This may not seem like a simplification since the new expression is longer than the original, but it will prove helpful when we consider ρ' to be small.

The conservation of momentum reads

$$\rho_0 \frac{D\vec{u}}{Dt} = -\nabla p_h(z) - \nabla p' - \rho_0 \bar{\rho}(z) g \hat{k} - \rho_0 \rho' g \hat{k}$$

where we have decomposed the pressure to have two parts, one of which is a function of z only and we have labeled $p_h(z)$. The other, like the perturbation density, is labeled with a "prime". If we assume that the stably stratified background state is not moving we get the following equation for p_h:

$$\frac{dp_h}{dz} = -\rho_0 \bar{\rho}(z) g.$$

This is a result often taught in first year physics courses and defines the so-called hydrostatic pressure, which is an expression for the weight of the fluid that overlies a fluid particle at z.

8.2 The idealized internal seiche: derivation

We can now write the conservation of momentum for only the "prime" or perturbation quantities:

$$\frac{D\vec{u}}{Dt} = -\frac{1}{\rho_0}\nabla p' - \rho' g\hat{k}.$$

In order to make theoretical progress we assume small perturbations so that the nonlinear terms can be dropped. Letting $\vec{u} = (u, w)$ we can get equations for the two velocity components, as well as for ρ' and the expression for conservation of mass. We will denote partial derivatives by subscripts for convenience so that

$$u_t = -\frac{1}{\rho_0} p_x \tag{8.5}$$

$$w_t = -\frac{1}{\rho_0} p_z - \rho' g \tag{8.6}$$

$$\rho'_t + w\bar{\rho}'(z) = 0 \tag{8.7}$$

$$u_x + w_z = 0 \tag{8.8}$$

Since we assumed the fluid is inviscid so that at $z = 0, H$ we have $w = 0$ and at $x = 0, L$ we have $u = 0$. Our strategy is to form a single equation for wave like disturbances. The divergence-free condition can be automatically satisfied if we take (u, w) to be defined from a streamfunction (recall the discussion in Section 4.4), ψ:

$$(u, w) = (\psi_z, -\psi_x).$$

We can now cross-differentiate and subtract the u and w equations to find that

$$(u_z - w_x)_t = \rho'_x g.$$

The density equation reads

$$\rho'_t = -w\frac{d\bar{\rho}}{dz}.$$

A tiny bit of algebra shows $u_z - w_x = \psi_{xx} + \psi_{zz}$ and taking a time derivative of the cross-differentiated momentum equation gives

$$\nabla^2 \psi_{tt} = \rho'_{xt} g.$$

Taking the x derivative of the density equation gives

$$\rho'_{tx} = \psi_{xx}\frac{d\bar{\rho}}{dz}.$$

We can now eliminate the perturbation density to get the single equation

$$\nabla^2 \psi_{tt} = \psi_{xx} g\frac{d\bar{\rho}}{dz}.$$

136 **CHAPTER 8** Modeling motion in the vertical

There is one more piece of business, in order to make our derivation complete. We define the so-called buoyancy (or Brunt-Vaisala) frequency $N(z)$ via the expression

$$N^2(z) = -g\frac{d\bar{\rho}}{dz}. \tag{8.9}$$

The governing equation then reads

$$\nabla^2\psi_{tt} + N^2(z)\psi_{xx} = 0 \tag{8.10}$$

with the boundary conditions

$$\psi_z = 0 \text{ at } x = 0, L \tag{8.11}$$
$$\psi_x = 0 \text{ at } z = 0, H. \tag{8.12}$$

To solve the equation we could use separation of variables, but we instead use a more transparent method to make some physically motivated assumptions. We wish to consider small amplitude waves, and sine and cosine thus make natural choices for the horizontal and temporal structure.

If we let

$$\psi = \sin(kx)\cos(\sigma t)\phi(z)$$

the boundary conditions at $x = 0, L$ are satisfied automatically provided

$$k = \frac{m\pi}{L}$$

for any integer m. We thus write our guess for a solution as

$$\psi_m = \sin(k_m x)\cos(\sigma_m t)\phi(z)$$

and substitution into the governing equation gives

$$-\sigma^2(\phi_{zz} - k^2\phi) - k^2 N^2(z) = 0$$

which can be simplified to read

$$\phi_{zz} + \left(\frac{N^2(z)}{\sigma^2} - 1\right)k^2\phi = 0$$

along with the boundary conditions $\phi = 0$ at $z = 0, H$. While this is a relatively simple second order differential equation eigenvalue problem (i.e. we solve for σ along with $\phi(z)$), for a general stratification it must be solved numerically.

However when the density varies linearly with z, or in other words $N^2(z) = N_0^2$, a constant, the problem can be solved analytically. The governing equation reads

$$\phi_{zz} + \left(\frac{N_0^2}{\sigma^2} - 1\right)k^2\phi = 0$$

and since the expression in the brackets is a constant, a reasonable guess is a solution of the form

$$\phi_n(z) = \sin(n\pi z/H)$$

for any integer n. The boundary conditions are immediately seen to be satisfied and the differential equation will be satisfied as well provided

$$\frac{N_0^2}{\sigma^2} - 1 > 0$$

which gives a restriction on the possible frequencies of oscillation

$$|\sigma| \leq N_0 \tag{8.13}$$

and provides one explanation for why $N(z)$ is called the buoyancy frequency. Substituting the trial solution gives an expression for the frequency,

$$-\frac{n^2\pi^2}{H^2} + \frac{k^2 N_0^2}{\sigma^2} - k^2 = 0$$

so that a complete solution in terms of the two indices m and n can be written as

$$\psi = \cos(\sigma_{mn}t) \sin(k_m x) \sin(m_n z) \tag{8.14}$$

$$\sigma_{mn} = \frac{N_0 k_m}{\sqrt{k_m^2 + m_n^2}} \tag{8.15}$$

$$k_m = \frac{m\pi x}{L} \tag{8.16}$$

$$m_n = \frac{n\pi x}{H}. \tag{8.17}$$

The solution is quite tidy and allows for a number of immediate conclusions.

1. The frequency takes discrete values (physicists would say it is "quantized").
2. Both k_m and m_n have a lowest value and so while $0 \leq |\sigma| < N_0$ how tight the two bounds are is not clear.
3. For most natural bodies of water $L \gg H$ so that the smallest k_m is much smaller than the smallest m_n.

The integer that sets m_n is often referred to as the mode number of the wave. While both mode-1 and mode-2 internal seiches have been reported in the literature, detecting motions that are purely composed of vertical modes higher than 2 is elusive (i.e. it is better to think of higher modes as a mathematical building block for coherent motions actually observed in nature). Similarly, a proper assessment of how many horizontal modes can be expected to be measured in Nature, would require a discussion of a concrete geographical situation. Nevertheless, low modes are often assumed to be far more common.

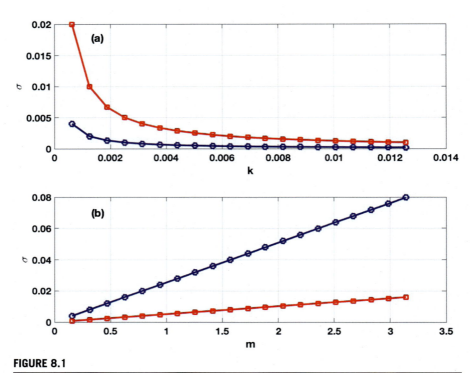

FIGURE 8.1

Frequencies of internal seiches in a linearly stratified "aquarium" lake. $H = 20$ m, $L = 5{,}000$ m, $\Delta\rho = 0.01$. (a) Frequency as a function of k for the first 100 horizontal modes with the first (blue) and fifth (red) vertical mode number, (b) frequency as a function of m for the first 100 vertical modes with the first (blue) and fifth (red) horizontal mode number.

Fig. 8.1 shows how the frequency varies as a function of horizontal (panel (a)) and vertical (panel (b)) wavenumber. Frequencies decrease as a function of the horizontal wavenumber and increases as a function of the vertical wavenumber. However, the key point to note is that when the stratification is linear, the spatial structure is always sinusoidal in both x and z.

8.3 The idealized internal seiche: transport

Now that we have derived a solution to the internal seiche problem let's try to do something useful with it. Before we do so, we should note a few caveats: first our solution is approximate because we linearized first, second even if we assume linearization is OK we should recall we made physical assumptions, namely that rotation is not important (which means we should restrict either the scale of our lake

8.3 The idealized internal seiche: transport   **139**

or the time on which we consider our solutions) and that viscosity can be ignored (which means we cannot meaningfully discuss the near bottom region).

For the interested reader, the above figure was generated by the script

```
constant_N_seiche_freq.m
```

If we wish to consider transport we have two choices. First, we could consider the transport of a passive tracer, like a dye in an experimental tank. The dye is measured at fixed locations in space, and this means it is referred to in fluid mechanics as an "Eulerian" variable, and its movement is referred to as **Eulerian transport**. Alternatively, we could consider the transport of particles, or what fluid mechanics practitioners call **Lagrangian transport**. The governing equations for a particle are pretty simple

$$\frac{d\vec{x}}{dt} = \vec{u}(x(t), z(t), t).$$

Since the velocity field is known analytically these can be solved by any number of standard numerical methods. The reader may wonder why we cannot solve the flow analytically. This is because the right hand side is evaluated at time dependent locations, so that the integral cannot be evaluated in terms of known functions. This is not a serious issue since a well-validated numerical method beats an overly complex analytical formula every time.

We consider a domain that is $H = 20$ m deep, $L = 5,000$ m long and a top to bottom dimensionless density difference of 0.01. We provide M-files to reproduce the figures we show below, though we note that the "live movie" script is far more fun to play with. The live movie maker has an added feature that the present location of the particles is shown in black and the particles at the previous output time are shown as smaller, white dots. We have chosen to not show this "particle history" for the images shown in the book text.

The sensible sanity test is to track a group of particles over one period of a small amplitude wave. The particles for all numerical experiments shown are initially located in a grid pattern for which $0.4L \leq x \leq 0.6L$ and $0.45H \leq z \leq 0.75H$. This choice was established based on trial and error, where the goal was to sample some of the time dependent horizontal shear. The relevant code is given in

```
constant_N_seiche_bookpics.m
```

and a live movie can be viewed by running the script

```
constant_N_seiche.m
```

The reader will have to change the settings, following the comments in the code, for the various figures shown in this chapter. While we concentrate on Lagrangian transport in the discussion, a code for Eulerian transport is also included for the readers, namely,

```
constant_N_seiche_tracer.m
```

140 CHAPTER 8 Modeling motion in the vertical

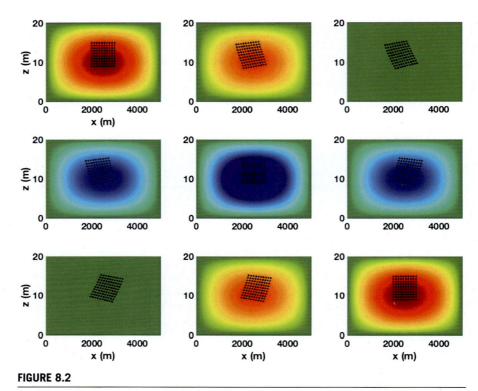

FIGURE 8.2

Particle locations over one period of a small amplitude mode-1 internal seiche. The streamfunction is shown shaded, with a saturation of ± 1.

Fig. 8.2 shows the particle locations at times that are 0 to 8 eighths of the mode-1 period. It can be seen that particles are advected primarily horizontally, and return to their initial position after each half period. This is essentially exactly what we would expect. The images, while very typical of how one visualizes internal waves, are misleading in terms of the scales of the problem. Were we to replot the result in 1-1 axes there would be nothing to see (the interested reader can try it by typing "axis equal" in MATLAB® after running one of the scripts). Given the small aspect ratio ($H/L = 0.004$), a better sense of particle motion is to focus in on the particle region.

In the previous case the dominant particle motion was in the horizontal. In Fig. 8.3 we consider a larger wave (7.5 times larger to be precise). It can be seen that particles are now advected in both the horizontal and vertical. They again return to an initial state after each half period. This may surprise the reader. After all it is a well known fact that surface water waves systematically transport particles in the direction of wave propagation through a phenomenon of Stokes drift, which has to do with the fact that particle paths are not exactly closed after one period. We note here that the seiche does not propagate, and hence the particle paths are essentially closed.

8.3 The idealized internal seiche: transport 141

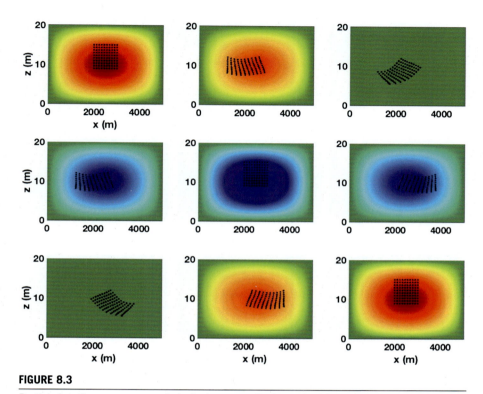

FIGURE 8.3

Particle locations over one period of a large amplitude mode-1 internal seiche. The streamfunction is shown shaded, with a saturation of ± 7.5.

For completeness, in Fig. 8.4 we show the particle transport by a mode-2 wave. The amplitude of the streamfunction is chosen to be the same as the mode-1 case shown in Fig. 8.2. In the mode-2 case the particle grid spans a region that experiences strong currents. This means particles experience vertical shear, with the lower portion of the original "box" being strongly advected in the horizontal. Nevertheless, the particles again return to their initial "box" pattern after each half period (note the mode-2 period is double that of the mode-1 seiche).

The interesting question is what happens when seiches of various modes are combined. Fig. 8.5 shows a case with the large mode-1 seiche with a small superimposed mode-2 seiche. Since there is no reason to expect the various modes to be synchronized in time, the mode-2 seiche is given a phase shift of 0.1. It can be seen that the particle motion is considerably more complex, and indeed particles do not return to the initial "box" pattern after one mode-1 period. This is perhaps not surprising. For this reason we integrated the particle paths for longer, recording their locations after 0 to 8 mode-1 periods of the mode-1 wave. Fig. 8.6 shows the result and it is very clear that the particles never return to their initial "box" and indeed after eight

142 CHAPTER 8 Modeling motion in the vertical

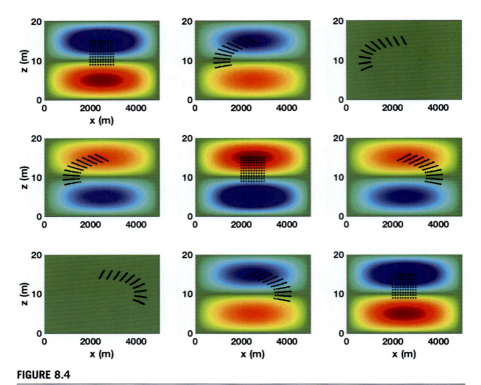

FIGURE 8.4

Particle locations over one period of a small amplitude mode-2 internal seiche. The streamfunction is shown shaded, with a saturation of ±1.

mode-1 periods the particles have a more or less chaotic distribution. The reader is invited to explore this further in one of the mini-projects at the end of the chapter.

Taken together these results suggest that while some particle motion is regular, it does not take much to break symmetry, and indeed we should expect patterns to break down quite quickly in Nature. This is precisely what makes observations of coherent transport in Nature so intriguing.

8.4 The internal seiche: arbitrary stratifications

The model of the previous two sections is very simplified. In this section we discuss the simplest possible extension. We accomplish this by keeping the theory linear but no longer demanding a specialized stratification. The price we pay for this is that the vertical structure must be solved for numerically. However, we can still accomplish a lot analytically. Moreover, when it comes time to do the numerics we get a hint

8.4 The internal seiche: arbitrary stratifications   143

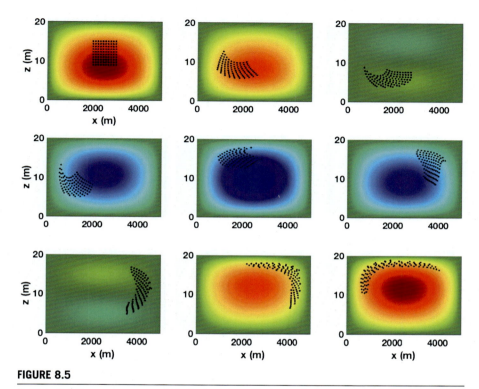

FIGURE 8.5

Particle locations over one period of a large amplitude mode-1 wave combined with a small amplitude mode-2 internal seiche. The streamfunction is shown shaded, with a saturation of ±7.5.

at what a full theory must accomplish while keeping both conceptual and numerical tractability.

We return to the last equation we derived prior to making the assumption of a linear stratification. This is Eq. (8.10) reproduced here for the reader's convenience:

$$\nabla^2 \psi_{tt} + N^2(z) \psi_{xx} = 0.$$

We demand that the stratification is constant so that $N^2(z) \geq 0$ but we assume that N^2 has some unspecified functional form. A well-studied particular example will be given below when we consider numerical examples. Motivated by the results of the previous section we guess a form for the streamfunction,

$$\psi(x, z, t) = \cos(\sigma t) \sin(kx) \phi(z).$$

The boundary conditions in the horizontal are unchanged and hence we get the same quantization condition for k, namely $k_m = m\pi/L$. Substitution and some simple al-

144 CHAPTER 8 Modeling motion in the vertical

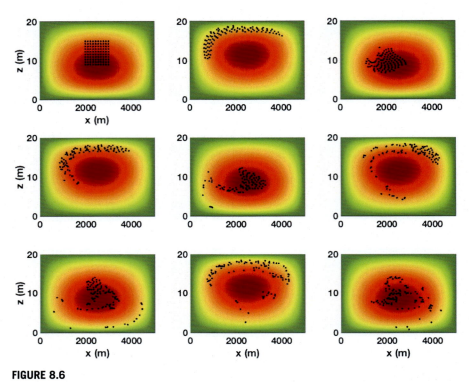

FIGURE 8.6

Particle locations over one period of a large amplitude mode-1 wave with combined with a small amplitude mode-2 internal seiche. The streamfunction is shown shaded, with a saturation of ±7.5.

gebra allows us to derive a second order differential equation for ϕ,

$$\phi_{zz} + \left(\frac{N^2(z)}{\sigma^2} - 1\right) k^2 \phi = 0. \tag{8.18}$$

Since the $w = 0$ boundary conditions at the top and bottom boundaries are unchanged, this equation is accompanied by the boundary conditions

$$\phi(0) = \phi(H) = 0$$

to make a second order differential eigenvalue problem.

This problem is relatively easy to solve via many different methods. In the code accompanying the book we use a so-called pseudospectral method that discretizes the derivative operation to create a matrix, and hence the second order ordinary differential equation is converted to a matrix generalized eigenvalue problem. We do this because MATLAB already has excellent routines for all sorts of linear algebra

8.4 The internal seiche: arbitrary stratifications   **145**

operations. This is an essential aspect of modern computing; use available routines that have been well-validated by others whenever possible.

Before we get into numerical methods let's consider the pathline equations again, but written in a way that explicitly accounts for the new, more general form of the solution

$$\frac{d\vec{x}}{dt} = \cos(\sigma t)[\sin(k_m x(t))\phi'(z(t)), k_m \cos(k_m x(t))\phi(z(t))]. \tag{8.19}$$

This expression nicely shows what is easy and what is hard about the more general version of the problem. We can still evaluate both $\cos(\sigma t)$, $\sin(k_m x(t))$ and $\cos(k_m x(t))$ analytically. However, after we solve a discretized problem we now only know $\phi(z)$ and $\phi'(z)$ at a discrete number of points, or in other words as a vector. That means that our pathline solver will have to also call on MATLAB's interpolation routines to get the appropriate velocities at the location a particle actually occupies at a given time. Fortunately, we only have to do this for the time varying locations in the vertical (i.e. $z(t)$).

For a general environmental fluid mechanics problem where the velocity $\vec{u}$ is only known on a model's grid we must perform an interpolation in either two or three dimensions (two for either an x-z model or a top down, x-y model). There are well set techniques for this, but it is an additional step of difficulty, as well as an additional computational cost (especially if we want to evaluate a large number of pathlines). Some numerical models have the means to evaluate particle paths as part of the model, but many solve for the velocity field first, and then solve for pathlines "offline". A number of open source packages are available with

```
https://oceanparcels.org/
```

being an example consistent with the spirit of this book. This saves lots of work computationally, but incurs a cost in terms of accuracy, and hence should be done carefully.

In many situations, stratifications can be considered to consist of a dominant pycnocline and a gradual background stratification. In freshwater lakes, stratification is typically due to differences in temperature. The code provided with this chapter allows the user to create various combinations of a pycnocline and background stratification. It then computes density using an approximation of the UNESCO equation of state, and uses it to solve the eigenvalue problem (8.18). An example of the stratification used for the examples in this section is shown in Fig. 8.7. The corresponding mode-1 and mode-2 vertical structure functions are shown in Fig. 8.8. It can be seen that while the background stratification changes the $N^2(z)$ profile it has only a minor effect on the vertical structure of both mode-1 and mode-2 waves. For this reason we will consider particle transport for the case with only a single pycnocline. The code to generate the above figures, and similar mode solutions for various lake relevant stratifications is given by

```
linear_lake.m
```

CHAPTER 8 Modeling motion in the vertical

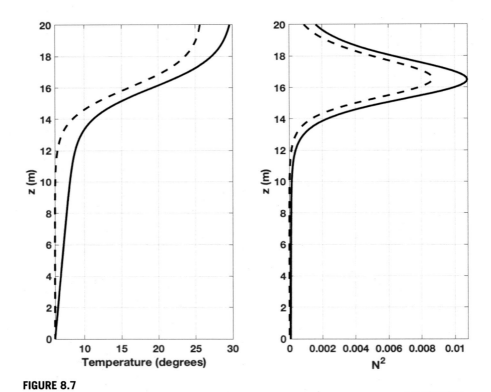

FIGURE 8.7

Sample temperature and $N^2(z)$ profiles for a single pycnocline and single pycnocline with an additional background stratification.

while the particle transport by an internal seiche for a variable $N^2(z)$ stratification, as shown in the figures discussed in the following paragraph, is given by

`variable_N_seiche_bookpics.m`

From Fig. 8.8 we can see that for mode-1 waves the primary effect of having a pycnocline is to shift the maximum from the center of the domain to roughly the pycnocline center (which is at 16 m in this case). The pycnocline center is the region of highest vertical shear. The particles are again distributed in a "box", though here we choose a somewhat larger box, with $0.35L \leq x \leq 0.65L$ and $0.55H \leq x \leq 0.95L$. Fig. 8.9 shows the particle locations at 0 to 8 eighths of the mode-1 period. The shear across the pycnocline is clearly evident with strongest horizontal currents occurring above the pycnocline. Like the constant N case, the particles return o the original position every half period.

Fig. 8.10 shows the corresponding evolution of the mode-2 seiche. Here the largest horizontal transport occurs near the pycnocline center, with weaker counter

8.5 The internal seiche: finite amplitude and wavetrains

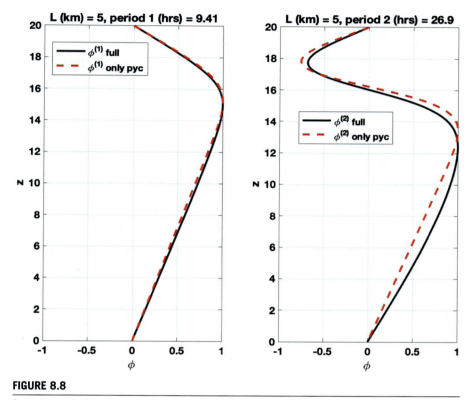

FIGURE 8.8

Sample mode-1 and mode-2 vertical structure functions for a single pycnocline and single pycnocline with an additional background stratification.

transport above and below the pycnocline. Once again the particles return to their initial "box" every half period. As reported in the panel titles in Fig. 8.8, it is no longer true that the mode-2 period is double that of the mode-1 period. In this case it is roughly 2.85 times larger.

8.5 The internal seiche: finite amplitude and wavetrains

While much can be learned from simplified theories, it is useful to situate the theory by performing more realistic simulations of the full stratified Navier Stokes equations. In this section we consider the evolution of an internal seiche in an "aquarium" or rectangular lake. The lake is chosen to have a depth of $H = 20$ m and a lateral extent of 5,000 m. The stratification is a quasi-two layer (i.e. single pycnocline) profile

148 CHAPTER 8 Modeling motion in the vertical

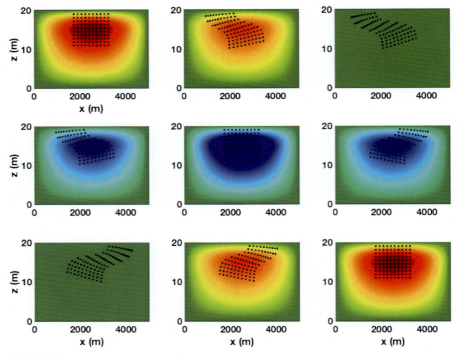

FIGURE 8.9

Particle locations over one period of a small amplitude mode-1 internal seiche with variable $N^2(z)$. The streamfunction is shown shaded, with a saturation of ± 1.

with the analytical form

$$\bar{\rho}(z) = 1 - \frac{\Delta\rho}{2} \tanh\left(\frac{z - z_0}{d}\right)$$

where $\delta\rho = 0.02$ sets the dimensionless density difference between the top and bottom, $z_0 = 17$ m is the pycnocline center, and $d = 1.5$ m is the pycnocline thickness. The simulation is initialized from rest by perturbing the density with a single cosine, or in mathematical form

$$\rho(\vec{x}, t = 0) = \bar{\rho}(z - 1.5\cos[\pi x/5000]).$$

Simulations were carried out with the stratified Navier Stokes solver SPINS, available from its User's guide page

```
https://wiki.math.uwaterloo.ca/fluidswiki/index.php?title=
SPINS_User_Guide
```

8.5 The internal seiche: finite amplitude and wavetrains   149

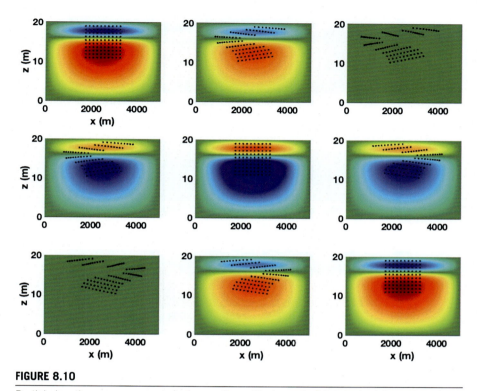

FIGURE 8.10

Particle locations over one period of a small amplitude mode-2 internal seiche with variable $N^2(z)$. The streamfunction is shown shaded, with a saturation of ± 1.

The horizontal resolution was approximately 1 m while the vertical resolution was approximately 5 cm. The discrepancy in resolution is necessary to resolve the stratification and the waves superimposed on it in detail. The model has spectral accuracy, implying that all motions with a scale of more than a few meters in the horizontal and about 20 centimeters in the vertical are well resolved. The simulation is initialized from a state of rest.

Fig. 8.11 shows the initial condition (upper panel) and the internal seiche after it has steepened ($t = 2.5$ hours) and begun forming a nonlinear wave train (middle panel, with detail shown in the lower panel). The nonlinear wave train takes the form of rank ordered bell-shaped solitary waves which are modulated so that the far upstream state transitions to a state with a depressed pycnocline that trails the wave train.

As Fig. 8.12 shows, the wavetrain continues to expand in horizontal extent as time goes on ($t = 3.25$ for this figure). The mode-1 wave train is trailed by a long mode-2 wave form (yellow arrow). The reason for the multi-mode response is that the initial conditions were not chosen to exactly project on one of the modes of linear

150 CHAPTER 8 Modeling motion in the vertical

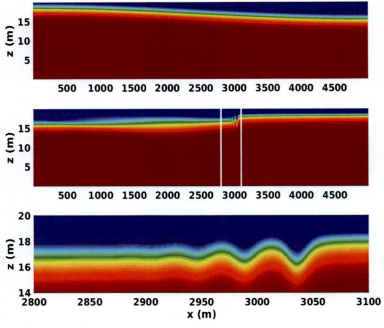

FIGURE 8.11

The shaded density field for the nonlinear stratified internal seiche. Upper panel – initial condition, middle panel – $t = 2.5$ hours, lower panel – $t = 2.5$ hours detail of region between white lines in the middle panel.

theory. Indeed these modes are a mathematical artifact, so that the sort of multi-mode response shown in Fig. 8.12 is far more representative of the response in a natural body of water due to a sudden wind event (e.g. a storm) than any aspect of linear theory.

A different perspective on the multi-mode response is offered by Fig. 8.13 which shows the horizontal component of the velocity with four isolines of density shown in white. The velocity field is capped at ± 0.25 m s^{-1}. For mode-1 the wave-induced velocities are opposite in sign at above and below the pycnocline, while for mode-2 the largest wave-induced velocities are in the deformed pycnocline. It is clear that the basin scale mode-1 and mode-2 responses are coupled. Fig. 8.14 shows a detailed picture of the nonlinear wave train. It can be seen the nonlinear wave train is responsible for the largest velocities below the pycnocline. The wave train is also responsible for the largest horizontal velocity gradients (both above and below the pycnocline).

When the wave train reaches the right boundary it begins to reflect off the solid boundary, and due its finite extent it begins to interact with itself. This leads to a complex wave breaking event that pushes even this rather finely resolved simulation past its design parameters. Nevertheless, in Fig. 8.15 we show the state at $t = 3.5$

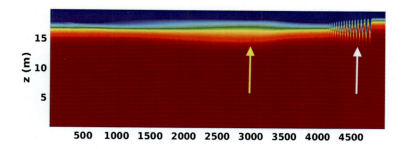

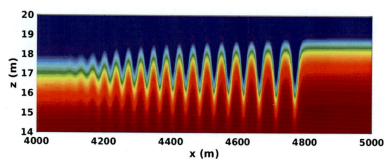

FIGURE 8.12

The shaded density field for the nonlinear stratified internal seiche. Upper panel – $t = 3.25$ hours, lower panel white arrow indicates mode-1 nonlinear wave train, yellow arrow indicates trailing long mode-2 wave, $t = 3.25$ hours detail of region near right wall.

hours. It can be seen that the largest amplitude leading waves that lead the wave train have emerged from the reflection process largely unscathed. They are, however trailed by a region over 300 m long in which a large level of small scale "turbulence" can be observed. The word turbulence in the previous sentence is in quotations, because this is a two-dimensional simulation for which the essential nature of turbulence (i.e. vortex stretching and tilting) is absent. Nevertheless, this simulation illustrates both how one can move beyond the layered models we had discussed in previous chapters, and just how far we are from resolving all the relevant scales of motion in an actual natural body of water. Indeed, it will be some time before a fully resolved model of even a small reservoir is available to the broad research community. Hence learning the tools and tricks of how to approximate successfully remains a vital step in the study of natural waters.

8.6 Nonlinearity due to biological behavior

In the previous section we discussed the effects of nonlinearity in terms of how the physics of the internal seiche evolved when the assumption of linearity was dropped.

CHAPTER 8 Modeling motion in the vertical

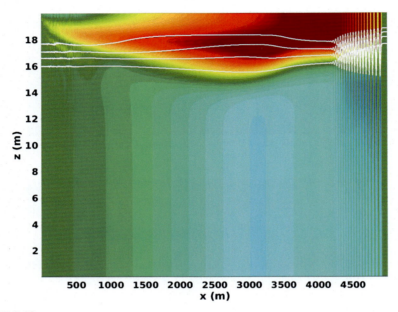

FIGURE 8.13

The shaded horizontal velocity field for the nonlinear stratified internal seiche at $t = 3.25$ hours with 4 isolines of density superimposed in white.

In this section we consider a very different nonlinearity, that due to more complex behavior on the part of the biological entities represented by the point particles of the first section of this chapter.

Biological entities do many things that Lagrangian particles do not: they may not follow the flow exactly (i.e. inertia and sinking), they may swim actively, or they may birth other particles. Let us examine two simple models and one more complex model of these features.

The simplest type of behavior associated with particles in natural waters is sinking. This is due to the fact that most living beings have a higher density than that of water (though the difference is often not very large). In a "full" theoretical treatment we would replace the Lagrangian particle equation

$$\frac{d\vec{x}}{dt} = \vec{u}(\vec{x}(t))$$

with a Newton's law based expression

$$m_p \frac{d^2\vec{x}}{dt^2} = \vec{F}_{drag} + \vec{F}_{buoyancy}$$

where the two terms on the right hand side represent the effects of the fluid flow on the particle, and the effect of the difference in mass of the particle from the mass

8.6 Nonlinearity due to biological behavior

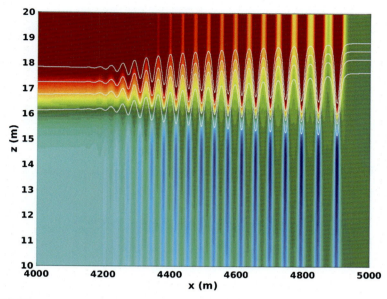

FIGURE 8.14

The shaded horizontal velocity field for the nonlinear stratified internal seiche at $t = 3.25$ hours with 4 isolines of density superimposed in white, in the region near the right wall.

of the volume of fluid it displaces (i.e. Archimedes" principle). The drag is often modeled, starting with Stokes" classical law, and possibly including corrections for higher Reynolds number, and particle shape. However, most studies of plankton are done with large scales in mind, meaning simpler models are thought to be better. The simplest way to include the main effect of sinking is to assume that the particle has achieved terminal velocity and hence that

$$\frac{d\vec{x}}{dt} = \vec{u}(\vec{x}(t)) - (0, -w_{sink}) \tag{8.20}$$

where w_{sink} represents a constant sinking velocity. The problem with a sinking velocity model is that something must bring the particles back up again, otherwise they simply collect at the bottom.

Most scientists invoke "turbulence" to keep particles from sinking out of the domain of interest. For large scale models (e.g. [1]) this is an unavoidable compromise, but even on small scales, very few models actually resolve turbulence relying instead on large values of eddy viscosity. Recent work ([2,4,6]) offers hope that this is changing. It is also possible that a particle could swim to counteract sinking and it is precisely this sort of model that we want to present here as a toy for the reader.

We consider the fluid motion to be given by the solution of Stokes' second (some books call it the third) problem: the motion of a constant density, viscous fluid which

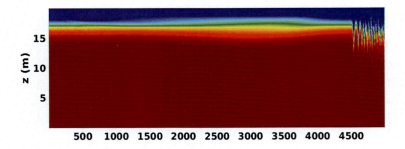

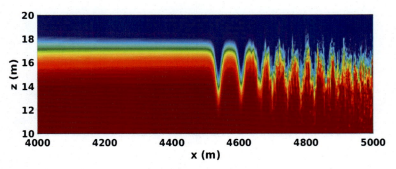

FIGURE 8.15

The shaded density field for the nonlinear stratified internal seiche. Upper panel – $t = 3.5$ hours, lower panel shows details of a 1 km long region near right wall.

occupies the upper half space overlying an oscillating plate. The advantage of this problem is that the solution is exact, yet quite simple. The motion of the plate is given by

$$u_{\text{plate}}(t) = U_0 \cos(\omega t). \tag{8.21}$$

The problem is simple, because the governing Navier-Stokes equations linearize geometrically in this case (or in other words without approximation) yielding the heat equation for the horizontal component of velocity

$$u_t = \nu u_{zz}. \tag{8.22}$$

This is easily solved ([5]) to give

$$u(z, t) = U_0 \cos(\omega t - mz) \exp(-mz), \tag{8.23}$$

while the vertical component of velocity is zero. Here

$$m = \sqrt{\frac{\omega}{2\nu}}. \tag{8.24}$$

8.6 Nonlinearity due to biological behavior

where ν is the kinematic viscosity. It can be seen that the vertical decay scale of motion ($L_d = 1/m$) and vertical period of oscillation ($L_o = 2\pi/m$) scale with the square root of viscosity, and the square root of the period of oscillation $T_p = 2\pi/\omega$. Both higher viscosity and slower oscillations of the plate increase penetration of the oscillations into the fluid. For the simulations shown below we fixed the viscosity to be that of seawater.

The swimming behavior of our particle (or plankton) must be triggered by something. While there are various interpretations in the literature ([8]) the simplest choice is to have swimming be triggered by shear. Since as the particle sinks and nears the plate, it is subjected to larger and larger shear variations, we can imagine a shear triggered response that leads it to swim "away", meaning upward.

With only one non-zero component of velocity the rate of strain tensor reduces to simple shear, given by

$$\frac{\partial u}{\partial z}(z, t) = -U_0 m \exp(-mz)\left[\sin(\omega t - mz) - \cos(\omega t - mz)\right]. \tag{8.25}$$

Using trigonometric identities and (8.24) this expression may be rewritten as

$$\frac{\partial u}{\partial z}(z, t) = -U_0 \sqrt{\frac{\omega}{\nu}} \exp(-mz) \sin\left(\omega t - mz - \frac{\pi}{4}\right), \tag{8.26}$$

The equations governing pathlines, modified by sinking and swimming reduce to

$$\frac{d\vec{x}}{dt} = [u(z(t), t), 0] + [0, w_{\text{sink}} + H(|u_z(z(t), t)| - u_z^{\text{critical}})w_{\text{swim}}], \tag{8.27}$$

Here w_{sink} is a constant, and we have modeled the shear triggered response using a Heaviside function: if the shear is higher than the critical value $u_z^{critical}$ the particles swim up. The code to generate the figures below is given by

```
simplest_particles.m
```

A rather surprising thing happens when we numerically integrate this system. The particle tends to a stable limit cycle in the approximate shape of a butterfly's wings. Fig. 8.16 shows four examples where we vary the critical shear by factors of two. It can be seen that the largest critical shear leads to the largest "butterfly" because the particle has reached deeper into the fluid, where the horizontal fluctuations are larger.

When the flow field is more complex, the simple shear is no longer sufficient to account for all shear, and the second invariant of the rate of strain tensor can be used to trigger swimming. The on-off type of swimming specified above may also need to be extended, and indeed a deterministic model may need to be replaced by a model that is stochastic. This can lead to models that rapidly become quite complicated. We leave these for the scientific literature (see [2,8] and the references therein). Instead we turn to a different biological aspect that is worth constructing models of,

156 CHAPTER 8 Modeling motion in the vertical

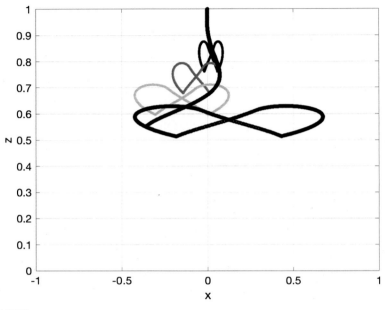

FIGURE 8.16

Particle paths for four sample shear triggered particles with values of critical shear changing. Colors range from the half base critical shear (thin black), base critical shear (dark gray), double base critical shear (light gray), and four times base critical shear (thick black).

namely birth and death of the particles. The code to generate the figures below is given by

   variable_N_seiche_wbirthdeath.m

We reconsider the variable $N^2(z)$ internal seiche, but this time from a total of 1,600 initial particles that occupy the entire domain, at each time step, we randomly choose one particle to give birth and one to die. The newly birthed particle has its initial condition specified to lie within a given distance (we choose $10^{-3}H$) of its parent particle. Fig. 8.17 shows the results of the simulation. It is apparent that even with what seems like a very low birth-death rate, the fact that particles are born at all very quickly leads to a disruption of the particle pattern from the grid. This is made even more striking if the reader compares Fig. 8.17 to Fig. 8.18 which uses the same code, but merely sets the number of particles that are born and die at each time step to be zero. It is amazing how quickly the birth-death model leads to a clumping of particles. Populations in Nature are often observed to be patchy, and birth-death provides a poignantly simple explanation for this fact ([9]).

8.6 Nonlinearity due to biological behavior

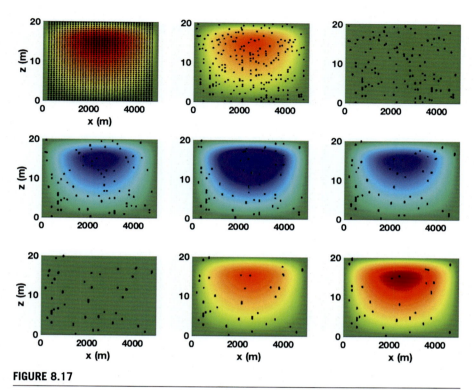

FIGURE 8.17

Particle locations for the model with birth and death, over one period of a small amplitude mode-1 internal seiche with variable $N^2(z)$. The streamfunction is shown shaded, with a saturation of ± 1.

A fairly significant branch of plankton-fluid coupling considers a very particular model. This model is called gyrotaxis and begins with the simple observation that most plankton are not spherical in shape. When one stops thinking of particles as infinitesimal, a statement of the balance of forces (the conservation of linear momentum) must be augmented with a balance of torques (the conservation of angular momentum). The theory has been worked out for various basic shapes, though for our arguments herein an ellipsoidal planktor (planktor is the term for an individual) will do. We assume the planktor has no variation in one angular variable and take the aspect ratio (or ratio of the longer axis and the shorter axis) to be labeled as q. In the simplest theory we now have a second vector quantity, namely the orientation of the planktor, typically labeled as $\vec{p}$. The vector sign distinguishes this from the pressure, though care should be taken.

For homogeneous particles a well developed theory of tumbling exists ([3,7]). For many types of plankton one further fact is believed to be important and that is the fact that one end of the ellipsoid is believed to be heavier than the other. This means that

158 CHAPTER 8 Modeling motion in the vertical

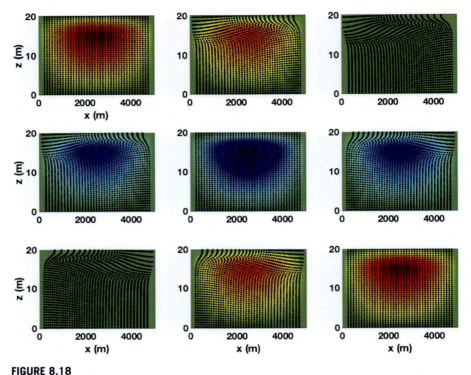

FIGURE 8.18

Particle locations for the model without birth and death, with the same initial conditions as the previous figure, over one period of a small amplitude mode-1 internal seiche with variable $N^2(z)$. The streamfunction is shown shaded, with a saturation of ± 1.

in a simple shear flow, the particle is subjected to two (often opposing) torques; one due to the effect of the shear and one due to gravity. This is schematized in Fig. 8.19.

There is currently considerable discussion on how to come up with tractable, effective, and computationally tractable models for gyrotaxis in turbulent fields. A good recent example is by De Lillo et al. ([2]) who give the equations of motion as

$$\vec{A} = \frac{D\vec{u}}{Dt} \qquad (8.28)$$

$$\frac{d\vec{p}}{dt} = \frac{1}{2v_0}[\vec{A} - (\vec{A} \cdot \vec{p})\vec{p}] + \frac{1}{2}\vec{\omega} \times \vec{p} \qquad (8.29)$$

$$\frac{d\vec{x}}{dt} = \vec{u} + v_c \vec{p}. \qquad (8.30)$$

In this model swimming with a speed v_c is held to happen in the direction of alignment, $\vec{p}$. The parameter v_0 is based on the distribution of mass, and De Lillo et al. give it as $v_0 = 3v/h$ where h is the distance of the center of mass to the geometric

8.6 Nonlinearity due to biological behavior

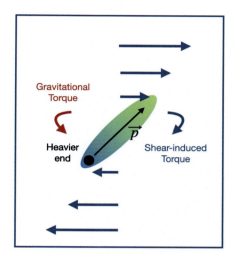

FIGURE 8.19

Diagram of gyrotaxis of an ellipsoidal particle in a shear flow.

center of the particle. The vorticity of fluid motion $\vec{\omega} = \nabla \times \vec{u}$ specifies the torques due to local shear.

The authors point out that this model, for which $\vec{A} = \vec{g} - \vec{a}$ modifies classical gyrotaxis which would take $\vec{A} = -g\hat{k}$. In the classical scenario the alignment stabilizes against gravity and swimming would occur only upwards.

The numerical costs of the above model, while not insignificant, are not overwhelming. At each time step one must:

1. Calculate the material derivative in order to get $\vec{A}$
2. Calculate the vorticity, $\vec{\omega} = \nabla \times \vec{u}$
3. Evolve $\vec{p}$
4. Evolve $\vec{x}$

The information that needs to be stored for each particle is doubled (i.e. two vectors as opposed to one) while two Eulerian vector fields must be computed and interpolated to the position of the particles.

The model as presented above is ideal for full numerical simulation. It is however far from clear how it should be simplified when simplified theories are used. For example, when we linearized to calculate solutions for the internal seiche, we effectively dropped the nonlinear terms in the material derivative. Should they thus be dropped in the expression for the acceleration (8.28)? This has not been answered in the literature, but we can take some preliminary steps here as a modeling exercise. The code to generate the figures below is given by

```
gyrotactic_stokes.m
```

160   CHAPTER 8 Modeling motion in the vertical

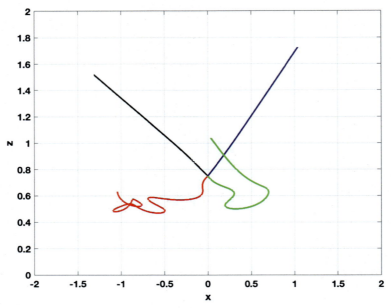

FIGURE 8.20

Paths of gyrotactic particles in the oscillating plate flow with four different initial orientations. Orientation is specified by angle, $\pi/4$ blue, $3\pi/4$ black, $-3\pi/4$ red, $-\pi/4$ green.

Let's reconsider the oscillating plate flow, and since it assumes a constant density for thhe fluid let's drop $\vec{g}$ from consideration. Indeed, it remains an open question whether the reduced gravitational acceleration should be used instead since it is the dfference in planktor and fluid densities that defines buoyancy for this system. In any event, since $\vec{u} = (u(z,t), 0, 0)$ only the x component of acceleration is non-zero, and this further implies that $\vec{a} \cdot \vec{p} = a^{(x)} p^{(x)}$ where the superscript denotes that these are the x components of the acceleration and the orientation vector. Moreover $a^{(x)} = u_t$ due to the geometric linearization of the Navier Stokes equations for this simple flow. The vorticity is also simpler, $\vec{\omega} = (0, u_z, 0)$ where the subscript denotes the partial derivative, and this implies that $\vec{\omega} \times \vec{p} = (u_z p^{(z)}, 0, -u_z p^{(x)})$. With these relations, the gyrotactic equations simplify considerably, and read

$$\vec{A} = \left(-\frac{\partial u}{\partial t}, 0, 0\right) \tag{8.31}$$

$$\frac{dp^{(x)}}{dt} = \frac{u_t}{2v_0}[1 - (p^{(x)})^2] + \frac{1}{2}p^{(z)}u_z \tag{8.32}$$

$$\frac{dp^{(z)}}{dt} = \frac{1}{2v_0}[-u_t p^{(x)} p^{(z)}] - \frac{1}{2}p^{(x)}u_z \tag{8.33}$$

8.6 Nonlinearity due to biological behavior

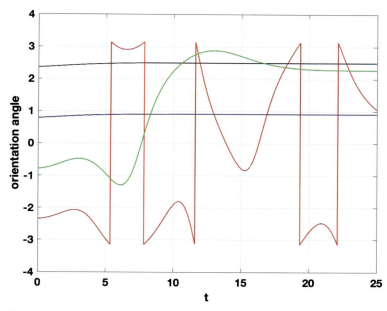

FIGURE 8.21

Orientation of gyrotactic particles in the oscillating plate flow with four different initial orientations versus time. Orientation is specified by angle, $\pi/4$ blue, $3\pi/4$ black, $-3\pi/4$ red, $-\pi/4$ green.

$$\frac{dx}{dt} = u + v_c p^{(x)}. \tag{8.34}$$

$$\frac{dz}{dt} = v_c p^{(z)}. \tag{8.35}$$

While these equations are quite complex, it is clear that z changes only due to the effect of swimming. From our toy swimming example we recall that upward directed swimming led to butterfly wing shaped orbits. We expect a more complex behavior here because not only is there swimming in both directions, the direction is time dependent.

Fig. 8.20 shows four paths that start at the same point, but have four different orientation directions, $n\pi/4$ where $n = 1, 3, -3, -1$, in blue, black, red, and green respectively. The vertical variable has been scaled by the starting point in the swimming particle model, for ease of comparison. It can be seen that the particles are initialized at 75% of the height of the simpler model presented above. The starting height was chosen by trial and error, and the reader should explore the code for themselves. From the figure, it is clear that the upward oriented particles are mostly unaffected by the current, while those oriented toward the bottom are profoundly affected, with orbits that change direction profoundly. The particle initially

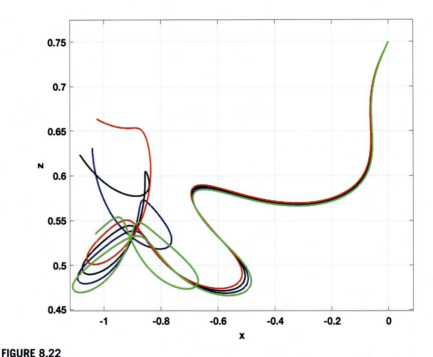

FIGURE 8.22

Paths of gyrotactic particles in the oscillating plate flow with four different initial orientations. Orientation is specified by angle, $-3\pi/4$ blue, $-3\pi/4 - 0.01$ black, $-3\pi/4 - 0.02$ red, $-3\pi/4 + 0.01$ green.

oriented in the $-\pi/4$ direction turns and swims upward, eventually escaping the high shear region. In contrast, the particle initially oriented in the $-3\pi/4$ direction has a complex path that remains in below the initial location for the full time of the simulation.

More information can be gained by considering the time dependence of the angle of orientation. This is shown in Fig. 8.21. Both initially upward oriented particles exhibit very little variation in orientation and certainly remain in the quadrant they were initialized in. The particle initially oriented in the $-\pi/4$ direction is seen to pass from the fourth to the first quadrant near $t = 8$, while the particle initially oriented in the $-3\pi/4$ direction crosses from the third to the second quadrant (i.e. the jumps between $-\pi$ and π) and indeed crosses into the first quadrant briefly between $14 < t < 16.5$.

The complexity of the particle path with an initial orientation of $-3\pi/4$ raises the question of sensitivity to small variations in initial conditions. While the remaining three initial orientations were found to have orbits that track over the times of our simulations, small changes of orientation, namely $-3\pi/4$, $-3\pi/4 - 0.01$, $-3\pi/4 - 0.02 - 3\pi/4 + 0.01$ track during the early evolution but rapidly depart when the angle

8.7 Mini-projects

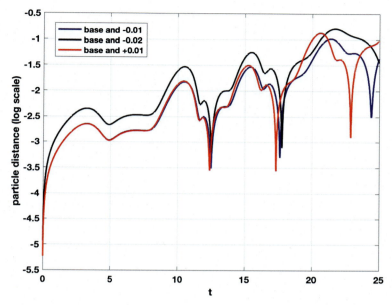

FIGURE 8.23

Inter-particle distance between the paths in the previous figure. All curves consider the difference from the $-3\pi/4$ case, with a perturbation of -0.01 blue, -0.02 black, $+0.01$ red.

switches to the first quadrant. The particle paths can be seen (with a magnified view) in Fig. 8.22. The inter-particle distance is a more quantitative measure of sensitivity to initial orientation and this is shown on a logarithmic scale in Fig. 8.23. There is an initial growth, albeit all distances are small at these early times. The shape of the distance curve is similar up to around $t = 18$ with a rapid divergence thereafter. This is also the time period for which the largest distances are observed.

We hope the reader agrees that given the simplicity of the underlying flow, these results are impressive (even if they are clearly of the "toy" variety).

8.7 Mini-projects

1. Use the code *variable_N_seiche.m* to solve for the internal seiche for a stratification of your own creation. Contrast the first three vertical and horizontal modes.
2. Use the code *variable_N_seiche.m* to solve for the internal seiche for a stratification of your own creation. Contrast the first three vertical and horizontal modes.
3. Modify the code *simplest_particles.m* so that the plate oscillates with two frequencies.

164 **CHAPTER 8** Modeling motion in the vertical

4. Use the code *variable_N_w_birthdeath.m* to solve for the dynamics of particles with birth and death for a mode-2 seiche. Use a stratification of your own creation.
5. Generalize the code *gyrotactic_stokes.m* to include more particles, and use it to study a cloud of particles starting at different heights.
6. Find examples in the literature of observations of internal seiches in lakes, and mode-2 seiches in particular. Is the evidence clear? How do images based on field data contrast with those based on simulations?

References

[1] João H. Bettencourt, et al., Effects of upwelling duration and phytoplankton growth regime on dissolved-oxygen levels in an idealized Iberian Peninsula upwelling system, Nonlinear Processes in Geophysics 27 (2) (2020) 277–294.
[2] Filippo De Lillo, et al., Turbulent fluid acceleration generates clusters of gyrotactic microorganisms, Physical Review Letters 112 (4) (2014) 044502.
[3] Jeffrey S. Guasto, Roberto Rusconi, Roman Stocker, Fluid mechanics of planktonic microorganisms, Annual Review of Fluid Mechanics 44 (2012) 373–400.
[4] Alice Jaccod, et al., Three-dimensional turbulence effects on plankton dynamics behind an obstacle, The European Physical Journal Plus 137 (2) (2022) 1–11.
[5] P.K. Kundu, I.M. Cohen, Fluid Mechanics, 4th ed., Elsevier Academic Press, 2008.
[6] Clotilde Le Quiniou, et al., Copepod swimming activity and turbulence intensity: study in the Agiturb turbulence generator system, The European Physical Journal Plus 137 (2) (2022) 1–14.
[7] T.J. Pedley, John O. Kessler, Hydrodynamic phenomena in suspensions of swimming microorganisms, Annual Review of Fluid Mechanics 24 (1) (1992) 313–358.
[8] Justin Shaw, Marek Stastna, A model for shear response in swimming plankton, Progress in Oceanography 151 (2017) 1–12.
[9] William R. Young, Anthony J. Roberts, Gordan Stuhne, Reproductive pair correlations and the clustering of organisms, Nature 412 (6844) (2001) 328–331.

CHAPTER

Modeling fluid transport processes with finite volume methods

9

CONTENTS

9.1	Beyond spectral methods	165
9.2	A short history of finite volume methods	166
9.3	Getting acquainted with finite volume methods in 1D	166
9.4	Insight from 1D energy analysis: the continuous problem	170
9.5	Insight from 1D energy analysis: the discrete problem	171
9.6	Exercises: building intuition	173
9.7	Implementation details in 1D	173
9.8	Doing it "the right way" in 2D	174
9.9	Towards harder problems	177
9.10	Well-balanced upwinding for the case of variable bottom bathymetry	179
9.11	Returning to the dispersive shallow water system	181
9.12	Towards "high-resolution" numerical schemes	183
9.13	Wrap-up	184
9.14	Mini-projects	184
References		185

9.1 Beyond spectral methods

We have now completed a tour of the modeling aspects of biological tracers influenced by evolving motions of fluids. Along the way, we have stressed the need to make simplifying assumptions that allowed for mathematical progress. At the same time, we have showcased the amazing progress one can make if one makes ready use of numerical methods. The numerical methods of the previous chapters often relied on the use of the FFT (i.e. Fourier spectral methods). This allowed for methods with minimal numerical dissipation, and at times some rather impressive graphics (Chapters 7 and 8, in particular).

The modeling of actual lakes, requires one to account for complex boundaries and as such FFT-based methods are not immediately useful. Historically, numerical models have been far more reliant on finite volume methods. In this chapter we present these in a relatively self-contained way, with minimal mathematics (though there is some need for basic calculus and linear algebra). This is intended as both background and a cautionary tale, that motivates the rather more advanced methods of Chapters 10 and 11.

Physics and Ecology in Fluids. https://doi.org/10.1016/B978-0-32-391244-0.00019-X
Copyright © 2023 Elsevier Inc. All rights reserved.

165

166 **CHAPTER 9** Finite volume methods

9.2 A short history of finite volume methods

The finite volume method encompasses a rich history, from the pioneering "rocket scientists" [5] to oceanographers [4], and even modern theoretical physicists [1]. Seemingly all dynamically-rich branches of science have benefited from their robustness, numerical stability, and (apparent) favorable computational performance. Due to this long history of adoption and study, one can even say that "they've already been done" (see [12]), and are mathematically well-understood thanks to those highly dedicated mathematicians who are well-versed in the theory of hyperbolic partial differential equations (PDEs). Why then, is the finite volume methodology being recounted in an environmental modeling textbook?

In a nutshell, a finite volume method says that "(naturally) conserved quantities ought to be *conserved* (in some finite, analogous way) on the computer that is simulating them." This property is desirable not only in forming the basis of a conceptually intuitive computer model, but because of the rather "nice" properties that occur due to this stipulation (see "benefits" in above paragraph, for a few examples). Furthermore, more sophisticated numerical methods tend to parrot this desire to "conserve things," e.g., energy and enstrophy [2], or leverage the finite volume method's framework as a "starting point" in some other way; see, for example, the high-resolution Discontinuous Galerkin method [10].

9.3 Getting acquainted with finite volume methods in 1D

Consider the simple 1D tracer transport equation (as seen previously in Eq. (3.2) with $\alpha = 0$), also known as the "linear advection equation"

$$\frac{\partial T}{\partial t} + c\frac{\partial T}{\partial x} = 0, \tag{9.1}$$

where $c > 0$ is constant, in the closed section of the real line $\Omega = [0, L]$. The finite volume method presumes the domain under consideration Ω may be subdivided into a finite number of cells (or elements) that can be regarded as control volumes for conserved quantities that, in the simplest case, are approximated as constants throughout the extent of each cell, although this choice is not a unique one (see [8] for high-order reconstruction schemes for advection). In the case of a simple line-segment domain Ω with length L we might be driven to choose a straightforward partitioning based on uniformly-sized sub-interval cells. That is, we can assume that the line segment Ω divided into N sub-segments then the center of the finite volume cell at index j is given by the discrete Cartesian coordinate

$$x_j = \left(j + \frac{1}{2}\right)\Delta x, \quad j = 0, \cdots, N - 1, \tag{9.2}$$

where $\Delta x = L/N$ is the length of each subinterval. A graphical depiction of this discretization approach is shown in Fig. 9.1.

9.3 Getting acquainted with finite volume methods in 1D

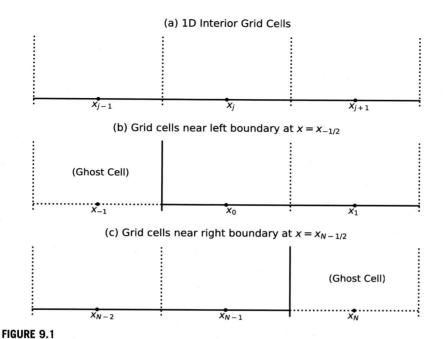

FIGURE 9.1

Graphical depiction of 1D grid cells lying in a line segment along the real line in the finite volume method. Ghost cells, which are used to impose boundary conditions at the first and last cell edges but are not actually a part of the computational grid, are also shown.

The end-goal of the finite volume method is to derive a discrete time-evolution recipe for the cell-centered average of the flow variable

$$\overline{T}_j = \frac{1}{\Delta x} \int_{x_{j-\frac{1}{2}}}^{x_{j+\frac{1}{2}}} T(x,t)\,dx. \tag{9.3}$$

To achieve this goal, we first integrate over cell j and apply the second fundamental theorem of calculus to the second term on the left-hand side, i.e.,

$$\frac{d}{dt}\int_{\Omega_j} T(x,t)\,dx + cT|_{x_{j-\frac{1}{2}}}^{x_{j+\frac{1}{2}}} dx = 0, \tag{9.4}$$

where we have brought the time-derivative outside the integral by the virtue of the fact that the cells" boundaries do not change with time. Using the definition of the cell average (9.3), and at the same writing the long-hand form of the bracketed expression we have

$$\Delta x \frac{d\overline{T}_{ij}}{dt} + \left[(\widetilde{cT})_{j+\frac{1}{2}} - (\widetilde{cT})_{j-\frac{1}{2}}\right] = 0 \tag{9.5}$$

168 CHAPTER 9 Finite volume methods

where we have used a subscript notation as a short-hand convention, and introduced ~'s to denote a "special value" since the function values at the cell-edges are non-unique because every cell has a left neighbor and a right neighbor that, in general, should be taken into account. Dividing by Δx, this equation can be written in the "flux-differencing" form

$$\frac{d\overline{T}_j}{dt} + \frac{\left(\widetilde{cT}\right)_{j+1/2} - \left(\widetilde{cT}\right)_{j-1/2}}{\Delta x} = 0. \tag{9.6}$$

To arrive at a time-evolution formula, we simply integrate from one discrete time level to the next $t_{n+1} = t_n + \Delta t$ and use the mean value theorem on the one-dimensional time integral. After some rearranging, we find:

$$\overline{T}_j^{n+1} = \overline{T}_j^n - \frac{\Delta t}{\Delta x}\left[\left(\widetilde{cT}\right)_{j+\frac{1}{2}}^{n+\alpha} - \left(\widetilde{cT}\right)_{j-\frac{1}{2}}^{n+\alpha}\right] \tag{9.7}$$

where $0 \leq \alpha \leq 1$ is an unknown number and is, in general, a fractional time-level that the bracketed expression is evaluated at. It is worth noticing that no approximations have been made in arriving at the formula (technically a "recurrence") (9.7), and it is formally an exact discrete time-evolution formula based on a conservation principle. Though interesting from a theoretical point of view, this fact offers us little help until we can make progress on answering some remaining practical questions:

1. What should α be chosen to be, if we cannot know its exact value?
2. The flux function $\left(\widetilde{cT}\right)_{j+1/2}$ is evaluated along cell edges that border on a discontinuous "jump" to the adjacent cell values. How do we specify a flux value at a discontinuity? And is there a unique, or at least appropriate, choice here to guarantee the desired "discrete conservation" property holds?

To answer question (1), we might consult an expert in numerical integration methods who would tell us that if $\alpha = 0.5$, we can expect a more accurate answer than if we choose $\alpha = 0$ or $\alpha = 1$. On the other hand, if we choose $\alpha = 0$ we recover a direct formula for $\overline{T}_j^{n+1}$ and it appears to be an attractive choice by this benefit alone. If we were to choose $\alpha = 1$, and inspect the equation we would notice a linear algebra problem to solve, and the conceptual simplicity and "light work-load" of the "fully explicit" $\alpha = 0$ scheme is gone. As it turns out, any scheme with $\alpha \neq 0$ results in a nonlinear algebraic problem to solve, generally speaking, and falls into the class of "implicit" numerical integration methods. For the moment, let us consider the $\alpha = 0$ scheme for its conceptual simplicity, and we'll explore improved schemes and the various trade-offs at a later point in Section 9.12).

To answer question (2) regarding flux choice, we must derive intuition from the physics of the problem at-hand in order to properly inform our choice of numerical method. Consider the case of three neighboring finite volume cells in one dimension as in Fig. 9.1(a), as a reductionist example.

9.3 Getting acquainted with finite volume methods in 1D

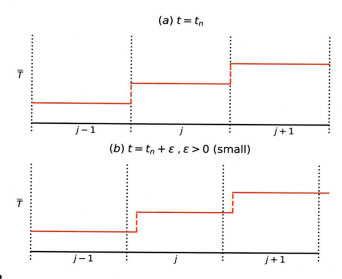

FIGURE 9.2

Cartoon diagram illustrating flow between neighboring finite volume cells for tracer field T and left-to-right velocity, $u > 0$. Panel (b) shows the discrete flow stage at a "fictitious time," after the solution at the edges has been reconstructed and evolved, but prior to finite volume averaging.

If the flow is from left-to-right, $u > 0$ as illustrated in Fig. 9.2, then immediately after being subject to the flow for a moment, our tracer T will have value from cell $j-1$ appear in cell j near its left-hand edge. Similarly, cell j's T-value will propagate into cell $j+1$. If the flow were, right-to-left, the situation would be reversed. In short, it appears that the most simple (and physically realistic) flux choice to "take information from where it is coming from." In the case of a left-to-right flow field, take fluxes from the left and in a right-to-left flow field, take fluxes from the right. Of course, it would be nice to generalize this statement somewhat, so let's think about the above explanation in terms of the unit-normal vector $\hat{\mathbf{n}} = \pm 1$ (in 1D), then we might write

$$(\widetilde{cT})_{j+\frac{1}{2}} = \begin{cases} (cT)_j, & c \cdot \hat{\mathbf{n}} \geq 0 \\ (cT)_{j+1}, & c \cdot \hat{\mathbf{n}} < 0 \end{cases} \quad (9.8)$$

and we would find an analogous expression for $(\widetilde{cT})_{j-\frac{1}{2}}$ by shifting indices via $j \leftarrow j-1$. This simple, yet intuitive, numerical flux choice is often called the "upwind flux," since it considers the source of information as upwind of a reference point in the case of a consistently blowing "wind" or "background flow."

170 **CHAPTER 9** Finite volume methods

9.4 **Insight from 1D energy analysis: the continuous problem**

All of the components of the basic finite volume scheme of "Godunov-type" have been laid out above in Section 9.3, yet more could be done to convince the skeptical reader that everything would work properly if we were to write it out as a computer program. To gain confidence in our approach, let's consider the simple case of a one-dimensional tracer $T(x)$ lying in the bounded interval $[0, L]$ that contains its end-points subject to a constant left-to-right flow $c > 0$. The governing equation is then simply

$$\frac{\partial T}{\partial t} + c\frac{\partial T}{\partial x} = 0 . \tag{9.9}$$

Although a general definition of "energy" for such a simple equation may not be common knowledge, we can probably convince ourselves that if T is a real-valued function in $[0, L]$, we can at least intuitively expect that the quantity T^2 is positive, and there are perhaps some conditions under which its integral (i.e., continuous sum) would be a constant, so let's try to derive an equation for the "total T^2." We begin by multiplying by T and integrating:

$$\int_0^L T\frac{\partial T}{\partial t} + Tc\frac{\partial T}{\partial x}\,dx = 0 . \tag{9.10}$$

Recognizing that we can move the new factors of T inside of derivative symbols via the chain rule, and that we can move the time-derivative outside, some rearranging gives

$$\frac{d}{dt}\int_0^L \frac{1}{2}T^2\,dx = -\int_0^L c\frac{\partial}{\partial x}\left(\frac{1}{2}T^2\right)\,dx . \tag{9.11}$$

Letting E be the integral on the left-hand side and simplifying the right-hand side gives

$$\frac{dE}{dt} = \frac{1}{2}cT^2\bigg|_{x=0} - \frac{1}{2}cT^2\bigg|_{x=L} , \tag{9.12}$$

which says that the change in total "energy" depends only on tracer that flows in through the left boundary, leading to a net increase in energy, minus the tracer that flows out through the right boundary, leading to a net decrease in energy.

In most practical situations, one expects the "entropy condition," or second law of thermodynamics, to hold over the domain of interest, i.e.,

$$\frac{dE}{dt} \leq 0 ,$$

that is "the total energy of a closed system cannot increase," which in our case implies that

$$T(0)^2 \leq T(L)^2 .$$

9.5 Insight from 1D energy analysis: the discrete problem

Given our intuitive definition of total energy for the continuous problem above, we may now pursue an analysis of the total energy in our spatially-discretized tracer function T_j. To that end, we recall that prior to integrating with respect to time, our finite volume method for the one-dimensional scalar transport equation at cell j reads

$$\frac{dT_j}{dt} = -\frac{\widetilde{(cT)}_{j+\frac{1}{2}} - \widetilde{(cT)}_{j-\frac{1}{2}}}{\Delta x}, \tag{9.13}$$

where we must choose a suitable formula for the numerical flux function $\widetilde{(cT)}_{j+\frac{1}{2}}$. Suppose for the sake of argument, that we are not yet sure whether the upwind flux is the best choice to make. More generally, we might take a weighted-averaged of neighboring cells, i.e.,

$$\widetilde{(cT)}_{j+\frac{1}{2}} = \beta cT_j + (1-\beta)cT_{j+1}, \tag{9.14}$$

where $0 \leq \beta \leq 1$ is a weight parameter, and it is presumed that $c > 0$, without loss of generality. It is self-apparent that we recover the upwind scheme when $\beta = 1$, and a simple-average is obtained when $\beta = \frac{1}{2}$. Substituting the weighted average for both $\widetilde{(cT)}_{j-\frac{1}{2}}$ and $\widetilde{(cT)}_{j+\frac{1}{2}}$ and enduring a bit of algebra, we find

$$\frac{dT_j}{dt} = -\frac{(1-\beta)cT_{j+1} - \beta cT_{j-1} - (1-2\beta)cT_j}{\Delta x}. \tag{9.15}$$

To form an energy equation, we multiply by T_j and sum over all cells, leading to

$$\frac{d}{dt} \sum_{j=0}^{N-1} \frac{1}{2}T_j^2 = -\frac{1}{\Delta x} \sum_{j=0}^{N-1} (1-\beta)cT_{j+1}T_j - \beta cT_{j-1}T_j - (1-2\beta)cT_j^2, \tag{9.16}$$

which can be written

$$\frac{d}{dt} \sum_{j=0}^{N-1} \frac{1}{2}T_j^2 = -\frac{1}{\Delta x} \sum_{j=0}^{N-1} (1-\beta)cT_{j+1}T_j - \beta cT_{j-1}T_j + \frac{1}{\Delta x}(1-2\beta)c \sum_{j=0}^{N-1} T_j^2. \tag{9.17}$$

Since all terms in the energy balance must be ≤ 0 to satisfy the entropy condition, it follows that we must have $\beta \geq \frac{1}{2}$ for a stable method to guarantee that the last term on the right-hand side is negative. Flux choices where $\beta \geq \frac{1}{2}$ are thus said to be, "upwind-biased" fluxes.

172 CHAPTER 9 Finite volume methods

To delve into this further, it is useful to investigate a few special cases of Eq. (9.17) for the weight parameter, β. For example, if $\beta = 1/2$ the equation can be rearranged as

$$\frac{d}{dt} \sum_{j=0}^{N-1} \frac{1}{2} T_j^2 = -\frac{1}{2\Delta x} \sum_{j=0}^{N-1} cT_{j+1}T_j - cT_{j-1}T_j ,$$

(9.18)

since the last term on the right-hand side vanishes in this case. If we write the first few terms, a pattern emerges, i.e.,

$$\frac{d}{dt} \sum_{j=0}^{N-1} \frac{1}{2} T_j^2 = -\frac{1}{2\Delta x} [(cT_1T_0 - cT_{-1}T_0) + (cT_2T_1 - cT_0T_1) + \dots$$

$$+ (cT_NcT_{N-1} - cT_{N-2}T_{N-1})]$$

$$= \frac{1}{2\Delta x} [cT_{-1}T_0 - cT_NT_{N-1}] ,$$

(9.19)

as the other terms cancel. We have arrived at a discrete analogue of Eq. (9.11), from which one can obtain simple conditions to guarantee non-increasing energy within a closed domain, depending on the choice of inflow/outflow boundary conditions. Notice that when the simple "pinned" homogeneous Dirichlet-type boundary conditions $T_{-1} = T_N = 0$ are applied, that energy is conserved (i.e., a constant in time). The discrete version of the entropy condition is recovered when the Neumann-type boundary conditions

$$\left. \frac{\partial T}{\partial x} \right|_{x=0} = 0 , \qquad \left. \frac{\partial T}{\partial x} \right|_{x=L} = 0 ,$$

(9.20)

are applied, which when spatially discretized amount to

$$T_{-1} = T_0 , \qquad T_N = T_{N-1} .$$

(9.21)

Substituting into the energy equation and applying the entropy inequality yields

$$\frac{d}{dt} \sum_{j=0}^{N-1} \frac{1}{2} T_j^2 = \frac{1}{2\Delta x} \left[cT_0^2 - cT_{N-1}^2 \right] \leq 0 ,$$

(9.22)

which, in practice, is equivalent to the continuous entropy condition (9.11).

9.6 Exercises: building intuition

1. (i) Show in the upwind case, where $\beta = 1$, that the spatially-discretized energy equation is given by

$$\frac{d}{dt} \sum_{j=0}^{N-1} \frac{1}{2} T_j^2 = -\frac{1}{2\Delta x} \left[cT_0^2 - cT_{N-1}^2 \right] - \frac{c}{\Delta x} \sum_{j=0}^{N-1} T_j^2 . \qquad (9.23)$$

Hint: Use the same sum reduction that was applied to simplify Eq. (9.18).
(ii) In a sentence, describe a physical interpretation of the "new term" on the right hand side in terms of the total energy of the system.

2. (i) Repeat exercise (1) for the "downwind" case, where $\alpha = 0$, and assign a physical interpretation to any new terms using a sentence or two.
(ii) Are there any problems with the downwind advection scheme? Why, or why not? Explain in terms of the physical principles discussed in this chapter.
(iii) Draw a cartoon diagram illustrating a region of warm temperature in cell $j - 1$ being carried rightwards by a consistently blowing wind with speed c towards cell j with a lower temperature. Explain in a few sentences how the finite volume method would update cell j in the upwind scheme's case. Repeat this exercise for the downwind advection scheme. Based on your sketches, which of these two schemes seems the most appropriate in practice? Explain why.

9.7 Implementation details in 1D

To implement the Godunov-type finite volume method on a computer does not require any sophisticated algorithms or data structures, and usually a high-school level understanding of programming (e.g., variables, loops, functions) is all that is required for background knowledge. For those interested in completeness, `tracer1d.py` is a python 3 script, which has a full implementation of the finite volume method in 1D. The very popular (almost ubiquitous) "numpy" and "matplotlib" python libraries are important for vectorized array-arithmetic and graphical plotting interfaces, respectively. For those familiar with MATLAB® (or Octave) syntax, these libraries can be thought of as "bridging the gap" between the purely mathematical MATLAB programming interface, and the more general purpose syntax of an "agile" scripting language like python.

For those less interested in the specifics of a language-specific implementation, we present the pseudocode of the algorithm below.

Graphical output from the finite volume code is shown in Fig. 9.3 where the initial condition is taken to be a simple cosine function. A keen-eyed reader will notice that in addition to propagating to the right, the initial profile is dampened over time somewhat as a result of the upwind scheme's inherent **numerical diffusion**, which vanishes in the limit of arbitrarily small grid cells (see Exercise 1).

174 **CHAPTER 9** Finite volume methods

Algorithm 3 Theoretical upwinding approach for the 1D tracer equation.

Require: parameters and initial conditions
    while n is less than number of time steps **do**
        COMPUTE The value of T in the left-hand ghost cell where $x = x_{-1}$.
        DEFINE Ghost-padded field T_{ghost}.
        for each cell $j \in \{0, 1, \cdots, N_x - 1\}$ **do**
            COMPUTE The upwind numerical flux $(cT)_{j-1/2} = (cT)_{j-1}$ (and $(cT)_{j+1/2} = (cT)_j$).
            COMPUTE The numerical solution at the new time-step via $T_j^{n+1} \leftarrow T_j^n - (\Delta t / \Delta x)\left[cT_j - cT_{j-1}\right]$
        end for
        $n \leftarrow n + 1$
    end while

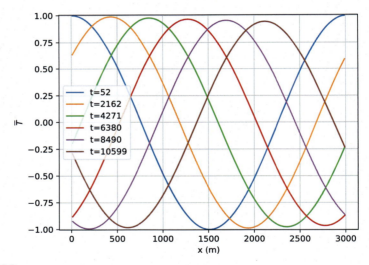

FIGURE 9.3

Upwind finite volume numerical solution to the 1D tracer equation for initial condition $T(x, 0) = \cos(2\pi x / L)$.

9.8 Doing it "the right way" in 2D

Consider the simple 2D tracer transport equation (as seen in Chapter 3)

$$\frac{\partial T}{\partial t} + \vec{u} \cdot \nabla T = 0. \quad (9.24)$$

The finite volume method presumes the domain under consideration may be subdivided into a finite number of cells (or elements) that can be regarded as control

9.8 Doing it "the right way" in 2D **175**

volumes for conserved quantities that are approximated as constants throughout the extent of each cell. In the case of a simple rectangular domain with length L and width H we might be driven to choose a straightforward partitioning based on uniformly-sized rectangular cells whose sides align with the and x- and y-axes. That is, we can assume that the x-axis is divided into sub-intervals and the y-axis is divided into sub-intervals then the center of the finite volume cell in row and column is given by the Cartesian coordinates

$$x_j = \left(j + \frac{1}{2} \right) \Delta x, \qquad j = 0, \cdots, N-1, \tag{9.25}$$

$$y_i = \left(i + \frac{1}{2} \right) \Delta y, \qquad i = 0, \cdots, M-1, \tag{9.26}$$

where Δx and Δy are the length and width of each rectangle, respectively. Here, we have adopted the convention that our indices i, j increase in the same direction as their corresponding coordinate (x or y) axis.

If the velocity field $\vec{u} = (u, v)$ is derived from streamfunction derivatives, i.e.,

$$u = -\frac{\partial \psi}{\partial y}, \tag{9.27}$$

$$v = \frac{\partial \psi}{\partial x}, \tag{9.28}$$

where ψ is a streamfunction, then via the product rule, the transport equation may be written in the more conservation-friendly form

$$\frac{\partial T}{\partial t} + \nabla \cdot \left(T \vec{u} \right) = 0. \tag{9.29}$$

Since $\nabla \cdot \vec{u} = 0$ by the definition of $\vec{u}$ in terms of the streamfunction. As in the 1D case, the end-goal of the finite volume method is to derive a discrete time-evolution recipe for the cell-centered average of the tracer variable

$$\overline{T}_{ij} = \frac{1}{\Delta x \Delta y} \iint_{\Omega_{ij}} T(x, y, t) \, dA. \tag{9.30}$$

To achieve this goal, we first integrate over cell (i, j) and apply Gauss' divergence theorem to the second term on the left-hand side, i.e.,

$$\frac{d}{dt} \iint_{\Omega_{ij}} T(x, y, t) \, dA + \oint_{\partial \Omega_{ij}} T \vec{u} \cdot \hat{n} \, ds = 0, \tag{9.31}$$

where we have brought the time-derivative outside the integral by the virtue of the fact that the cell boundary, is not time-dependent. Using the definition of the cell average (7.17), and at the same time breaking up the closed line integral into manageable 1D

176 **CHAPTER 9** Finite volume methods

pieces along the four edges of the rectangle, we have

$$\Delta x \Delta y \frac{d\overline{T}}{dt} + \left[\int_{x_{j-1/2}}^{x_{j+1/2}} -Tv\,|_{y=y_{i-1/2}}\, dx + \int_{y_{i-1/2}}^{y_{i+1/2}} Tu\,|_{x=x_{j+1/2}}\, dy \right.$$
$$\left. + \int_{x_{j+1/2}}^{x_{j-1/2}} Tv\,|_{y=y_{i+1/2}}\, dx + \int_{y_{i+1/2}}^{y_{i-1/2}} -Tu\,|_{x=x_{j-1/2}}\, dy \right] = 0. \quad (9.32)$$

This equation simplifies to

$$\Delta x \Delta y \frac{d\overline{T}_{ij}}{dt} + \left[(\widetilde{Tu})_{ij+1/2} - (\widetilde{Tu})_{ij-1/2} \right] \Delta y$$
$$+ \left[(\widetilde{Tv})_{i+1/2\,j} - (\widetilde{Tv})_{i-1/2\,j} \right] \Delta x = 0, \quad (9.33)$$

where we invoke the one-dimensional mean-value theorem in integral form. The variables with a superimposed tilde are evaluated at some unknown point along each line segment that is later assumed to be the midpoint, as an approximation. Dividing by $\Delta x \Delta y$, this equation can be written in the flux-differencing form

$$\frac{d\overline{T}_{ij}}{dt} + \frac{(\widetilde{Tu})_{ij+1/2} - (\widetilde{Tu})_{ij-1/2}}{\Delta x} + \frac{(\widetilde{Tv})_{i+1/2\,j} - (\widetilde{Tv})_{i-1/2\,j}}{\Delta y} = 0, \quad (9.34)$$

which is sometimes called a 'dimensional splitting' scheme since the motion in the x- and y-directions may be grouped and evolved independently.

To finalize the scheme, it remains to introduce discrete (or step-wise) time. A quick calculation shows that integrating from $t = t_n$ to $t = t_n + \Delta t$ is identical to how we derived the final scheme in the 1D case. After choosing the fully-explicit scheme, this operation yields:

$$\overline{T}_{ij}^{n+1} = \overline{T}_{ij}^{n+1} - \Delta t \left[\frac{(\widetilde{Tu})_{ij+1/2}^{n} - (\widetilde{Tu})_{ij-1/2}^{n}}{\Delta x} + \frac{(\widetilde{Tv})_{i+1/2\,j}^{n} - (\widetilde{Tv})_{i-1/2\,j}^{n}}{\Delta y} \right].$$
$$(9.35)$$

The implementation details for the finite volume method applied to the 2D tracer equation as discussed in this section are shown in Algorithm 4. Since "dimensional splitting" is utilized, the 1D upwind flux is simply applied in each direction separately. For those less interested in reading the full Python code, we list the generic algorithm in pseudocode here.

Graphical outputs from the `tracer2d.py` script are shown in Fig. 9.4, illustrating the chosen fluid velocity field $\vec{u} = (u(y), 0)$, where

$$u(y) = 0.2 \tanh \left(\frac{y - 1500}{300} \right),$$

Algorithm 4 Theoretical upwinding approach for the 2D tracer equation.

Require: parameters, initial conditions, and velocity field
   while n is less than number of time steps **do**
      COMPUTE The value of T, u, and v in the ghost cells where $x = x_{-1}, x = x_{N_x}$,
      $y = y_{-1}, y = y_{N_y}$.
      DEFINE Ghost-padded fields T_{ghost} (each time-step), u_{ghost} and v_{ghost} (only
      the first time-step).
      for each cell i, j **do**
         COMPUTE The upwind numerical fluxes $F_{ij-1/2}$, $F_{ij+1/2}$, $G_{i-1/2j}$,
         $G_{i+1/2j}$.
         COMPUTE The numerical solution at the new time-step via $T_j^{n+1} \leftarrow T_j^n -$
         $(\Delta t / \Delta x) [F_{ij+1/2} - F_{ij-1/2}] - (\Delta t / \Delta y) [G_{i+1/2j} - G_{i-1/2j}]$
      end for
      $n \leftarrow n + 1$
   end while

---

the initial datum for the tracer

$$T(x, y, 0) = \exp\left(-\left[\frac{x - 1500}{150}\right]^2\right) + \exp\left(-\left[\frac{x - 2250}{150}\right]^2\right)$$
$$+ \exp\left(-\left[\frac{x - 750}{150}\right]^2\right),$$

and the evolution of the field for two later output times. As in the simple 1D case, we note that tracers are simply translated for uniform flow. In the region where the shear u_y is appreciable, we find more interesting results where the tracer is stretched and spread along a narrow rotated elliptical region across the center of the domain.

9.9 Towards harder problems

In this chapter so far, we have covered transport problems where the flow moves in a prescribed (or known) direction. Consequently, the direction in which to apply the "upwind" scheme has always been well-known as well. It turns out that the upwind direction is not always the same as the direction of the velocity field. More generally, the proper direction for the upwind scheme must account for the direction in which *information propagates*, which may be significantly different than the local velocity direction in the case of wave propagation.

178 CHAPTER 9 Finite volume methods

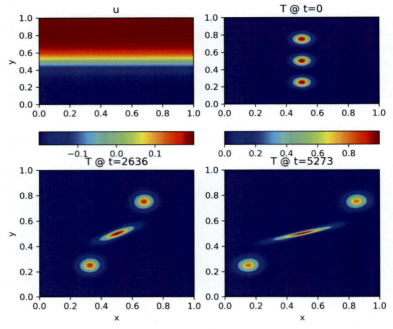

FIGURE 9.4

Output from the `tracer2d.py` script: A triple-Gaussian tracer field subjected to shear flow of the form $\vec{u} = (u(y), 0)$. *Top*: u velocity field component (left), initial tracer distribution $T(x, y, 0)$ (right). *Bottom*: Tracer distribution at $t = 2636$ s (left) and $t = 5273$ s (right).

A "non-diagonal" example

As a prototype problem, consider the one-dimensional shallow water system over a flat bottom bathymetry in the conservation form

$$h_t + (hu)_x = 0, \tag{9.36}$$

$$(hu)_t + \left(hu^2 + \frac{1}{2}gh^2\right)_x = 0. \tag{9.37}$$

Here h is the total layer depth, or $H + \eta$ in our previous notation. If we carefully apply the chain rule, we can put this system of two equations into what mathematicians refer to as **quasilinear form**

$$\begin{pmatrix} h \\ hu \end{pmatrix}_t + \begin{pmatrix} 0 & 1 \\ gh - u^2 & 2u \end{pmatrix} \begin{pmatrix} h \\ hu \end{pmatrix}_x = 0. \tag{9.38}$$

where the matrix appearing in this equation is called the **flux Jacobian** matrix, i.e.,

$$A = \frac{\partial \vec{F}(h, hu)}{\partial (h, hu)} \qquad \text{with} \qquad \vec{F} = \left(hu, hu^2 + \frac{1}{2}gh^2 \right)^{\mathsf{T}}. \tag{9.39}$$

To make progress from here, we observe that if the 2×2 matrix appearing in this equation were completely diagonal, then we would have a decoupled system and each equation could be solved independently as its own transport problem. Recall that from linear algebra, in order to diagonalize a matrix we require the eigenvalues and eigenvectors of the matrix to form the general decomposition

$$A = R \Lambda R^{-1}. \tag{9.40}$$

Here, A would be the 2×2 matrix from the quasi-linear problem, R would be the matrix consisting of A's eigenvectors as columns, and Λ would be the diagonal matrix with the corresponding eigenvalues of A as entries. One can show that in the quasi-linear case, we find

$$R = \begin{pmatrix} 1 & 1 \\ u - \sqrt{gh} & u + \sqrt{gh} \end{pmatrix}, \qquad \Lambda = \begin{pmatrix} u - \sqrt{gh} & 0 \\ 0 & u + \sqrt{gh} \end{pmatrix}. \tag{9.41}$$

In principle, we could go so far as to completely transform the problem using the approach described in the paragraph above, thereby finding a system of PDEs in terms of new flow variables. However, it turns out that projecting our solution vector (or its flux vector) onto the eigenvectors is sufficient to find the appropriate upwind numerical solution! For the interested reader, the theory for this projection is explained in full detail in Section 10.10 for the case of the linearized 1D shallow water system, where the underlying procedure is the same as for the quasilinear case. For the more practically minded reader, we illustrate the decomposition procedure in pseudocode algorithm (Algorithm 5) below.

9.10 Well-balanced upwinding for the case of variable bottom bathymetry

Suppose that the water depth may be decomposed at $h(x, t) = \eta(x, t) + H(x)$, where $H(x)$ is an undisturbed water depth that is constant in time and carries the spatial variations in bottom bathymetry. A common problem with numerical solutions to the shallow water system is that rapid variations in $H(x)$ can drive unphysical flow as a numerical artifact of a non-robust transport scheme, especially in situations where bottom bathymetry variations are poorly resolved. To see this, we write down the governing equations for the case where the pressure gradient $\eta_x \neq h_x$ due to a non-flat bottom. We can, however, use $\eta_x = h_x - H_x$ and rearrange the momentum equation

180 CHAPTER 9 Finite volume methods

Algorithm 5 Theoretical upwinding approach for the 1D nonlinear shallow water system.

Require: parameters and initial conditions
   while n is less than number of time steps **do**
      COMPUTE The values of h and hu in ghost cells x_{-1} and x_{N_x}.
      DEFINE Ghost-padded fields h_{ghost} and $(hu)_{ghost}$.
      COMPUTE The jump in the state vector $\vec{\delta q} = \left[h_{ghost,R} - h_{ghost,L}, \right.$
      $\left. (hu)_{ghost,R} - (hu)_{ghost,L} \right]^{\mathsf{T}}$.
      for each $j \in \{0, 1, \cdots, N_x - 1\}$ **do**
         COMPUTE The entries of the 2×2 matrix of eigenvectors.
         SOLVE The 2×2 matrix problem $R\vec{\alpha}_j = \vec{\delta q}_j$ for the projection coefficients
         $\vec{\alpha}_j = \left[\alpha_{j,1}, \alpha_{j,2} \right]^{\mathsf{T}}$.
         COMPUTE The leftward wave $\mathcal{W}^L = \alpha_1 \vec{r}_1$ and rightward wave $\mathcal{W}^R = \alpha_2 \vec{r}_2$,
         where $\vec{r}_i$ is column i of matrix R.
         COMPUTE The numerical solution at the new time-step via $h_j^{n+1} \leftarrow h_j^n -$
         $(\Delta t/\Delta x) \left[\mathcal{W}_{j+1,1}^L + \mathcal{W}_{j,1}^R \right]$ and $(hu)_j^{n+1} \leftarrow (hu)_j^n - (\Delta t/\Delta x) \left[\mathcal{W}_{j+1,2}^L + \right.$
         $\left. \mathcal{W}_{j,2}^R \right]$
      end for
      $n \leftarrow n + 1$
   end while

---

so that the PDE system reads

$$h_t + (hu)_x = 0, \tag{9.42}$$

$$(hu)_t + \left(hu^2 + \frac{1}{2}gh^2 \right)_x = ghH_x. \tag{9.43}$$

We expect a well-balanced scheme to preserve states of a perfectly flat free-surface, $\eta = 0$, and "no flow," $u = 0$, as constant in time, even over a non-flat bottom. That is, no non-physical flow is generated due to the advection scheme in the presence of the source term on the right-hand side. Taking $u = 0$ and taking time-derivatives to be 0, we reach the necessary condition

$$\left(\frac{1}{2}gh^2 \right)_x = ghH_x. \tag{9.44}$$

Recalling that the *de facto* finite volume method averages over each cell, we require

$$F_{j+1/2} - F_{j-1/2} = g \int_{j-\Delta x/2}^{j+\Delta x/2} hH_x \, dx. \tag{9.45}$$

To make progress with the integral on the right-hand side, we assume we can replace h with a unique value at the finite volume cell edges. In other words, at $x_{j+1/2}$, suppose h takes some unique value $h = h^{\dagger}_{j+1/2}$, then

$$F_{j+1/2} - F_{j-1/2} = gh^{\dagger}_{j+1/2}H_{j+1/2} - gh^{\dagger}_{j-1/2}H_{j-1/2}. \qquad (9.46)$$

It follows then that a well-balanced scheme results if we augment the flux function via

$$F_{j+1/2} \leftarrow F_{j+1/2} - gh^{\dagger}_{j+1/2}H_{j+1/2}. \qquad (9.47)$$

Once the above substitution for the numerical fluxes has been made for $F_{j-1/2}$ and $F_{j+1/2}$, the flux jump at each cell edge can then be projected onto the eigenvectors of the matrix A from the above section in what has been termed the F-wave approach [13]. Algorithm 5 requires only slight adjustments to arrive at the well-balanced scheme. Rather than repeat a very similar algorithm listing here, we encourage the reader to review the python script entitled `sw1d_fwave.py` for the details.

Exercise: Show that Eq. (9.46) is perfectly balanced when (9.47) is applied along with the uniqueness choice $h^{\dagger}_{j+1/2} = \frac{1}{2}(h_{j+1} + h_j)$. Hint: you may assume that information is propagating from left-to-right, without loss of generality so that $H_{j+1/2} = H_j$.

9.11 Returning to the dispersive shallow water system

So far, in this chapter we have only covered what mathematicians would call "fully hyperbolic" problems with and without source terms (e.g., the bathymetry term). We now expand our scope to weakly non-hydrostatic (dispersive) shallow water equations in 1D, that can be said to be "nearly hyperbolic," since the weak linear wave dispersion term may be regarded as only a small correction to the fully hydrostatic problem. Recall we simulated these using spectral methods in Chapter 6.

The PDE system is given by

$$h_t + (hu)_x = 0, \qquad (9.48)$$

$$(hu)_t + \left(hu^2 + \frac{1}{2}gh^2\right)_x = ghH_x + \frac{H^2}{6}(hu)_{xxt}, \qquad (9.49)$$

where only the last term appearing in the momentum equation is "new." To solve this problem, we follow the approach from earlier chapters where both time derivatives are discretized, and a linear algebra problem arises. The full problem can then be "split" up to resemble one time-step of the fully hyperbolic system, plus the solution of linear algebra matrix problem corresponding to what mathematicians call an "elliptic" or Poisson-type problem. The details of the implementation get somewhat involved, and we refer the reader to the source code file named `sw_1d_nonhydro_fv.py`.

182 CHAPTER 9 Finite volume methods

The solution procedure has been validated using a grid-convergence study, illustrated in Figs. 9.5 and 9.6.

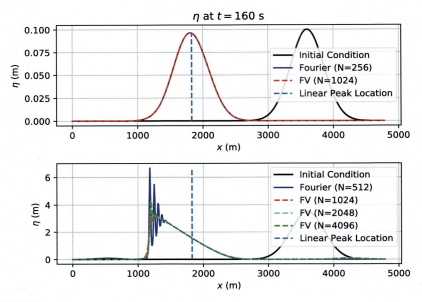

FIGURE 9.5

Top: The evolution of a small-amplitude Gaussian bump initialized to propagate from right-to-left as described by the finite volume (FV) method with $N = 1024$ grid cells, and the Fourier method with $N = 256$ grid points. The vertical dashed line appearing in both panels represents the wave peak as predicted by linear theory $x^* = x_0 - \sqrt{gH}t^*$, with $t^* = 160$ s.
Bottom: Like the top panel, but the initial condition's amplitude has been increased from 0.1 m to 3.5 m, and with successively doubled FV grids of size $N = 1024$, 2048, and 4096.

The problem consists of taking a Gaussian-shaped free surface elevation profile and choosing the velocity function u to correspond to a linear wave propagating to the left. Results are shown for two cases of the initial maximum wave height, first taken to be $\eta_{max} = 0.1$ m, then to be $\eta_{max} = 3.5$ m. The physical results are rather dull in the 0.1 m case, but more vividly interesting in the 3.5 m case. The numerical results point to the fact that the resolution capabilities of the Fourier method are far superior to the Finite Volume method for this particular problem, even for a comparatively small (2–4 times smaller) set of grid points. The reader must be careful to not interpret this result as a "blanket endorsement" of the Fourier method over the FV method. Rather, we leave it as a thought-exercise to consider in what problem scenarios and situations one might be right in choosing a certain method over another. Always remember: *"use the right tool for the job!"*

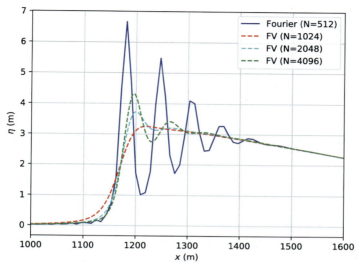

FIGURE 9.6

A repeat view of the bottom panel of Fig. 9.5, re-centered to show the details of the wave-trains more closely.

9.12 Towards "high-resolution" numerical schemes

The dispersive wave propagation problem illustrated above points to an issue with the standard "Godunov-type" finite volume method when applied to the problem of modeling dispersive short waves. In particular, we have shown that a disproportionately high number of grid cells are required to satisfactorily solve the problem when compared to the spectral (Fourier) method.

This particular short-coming has been highlighted in the ocean modeling literature for the situation of internal solitary wave propagation modeled using oceanic general circulation models (in "non-hydrostatic" mode). See [14] for an enlightening discussion. To overcome this problem for non-periodic geometries and in 2D/3D with sophisticated geometry (e.g., coastlines), our numerical modeling instinct says we need to find some other, "better" (or perhaps more "flexible"), approach. From this point, the literature typically branches off into two directions:

1. **FV approach:** Seek high-order corrections to the Godunov-type upwinding scheme, to limit the amount of numerical diffusion, thereby allowing for a method with better resolution characteristics and (faster) "convergence speeds." See for instance Durran's [8] discussion of anti-diffusion corrected third-order upwinding, or the second-order accurate MUSCL-Hancock scheme, as was used successfully by de la Fuente et al. [7] to model dispersive effects in closed circular basins. Geometric flexibility can be achieved by limiting which cells are included from

184 CHAPTER 9 Finite volume methods

the rectangular domain, using geometric mapping methods, or by using so-called "cut-cells" [3] to represent the boundary.

2. FEM approach: Use a (higher-order than linear) polynomial finite *element* method (FEM) capable of handling information that propagates with a preferred direction. These include the class of "Discontinuous Galerkin" methods [6,10] as well as the class of (continuous) streamline upwind Petrov Galerkin (SUPG) methods [11]. In the case of all FEM methods, finite element meshing software (free or otherwise licensed) are readily available to help supply accurate representations of physical boundaries. See, for example, the Gmsh mesh tool [9]. It is worth noting that simple finite volume methods can be made to work with unstructured grids also, but the complexity "explodes" when adopting higher-order advective corrections since information from more and more nearby grid cells is required.

In the following chapter, we introduce the Discontinuous Galerkin FEM (DG-FEM) approach, usually with polynomial orders between 4 and 8, to allow for a method both with enhanced resolution and convergence characteristics as well as (finite representation) geometric flexibility in two dimensions. The DG-FEM may be thought of as an extension of the FV method – by adding more points to a finite volume cell, corresponding to polynomial interpolation nodes. It may also be thought of as a domain-decomposition approach applied to a polynomial (Legendre) pseudospectral method, where continuity between subdomains is enforced in a "weak" sense, leveraging suitable FV-style numerical flux functions.

9.13 Wrap-up

We have covered considerable ground in this chapter, pulling in tools from calculus, physics, and computer science. Mini-projects below offer the reader an opportunity to build an experiential knowledge foundation on what they've read. Often, "the school of doing it" can be a lot more helpful than the school of just "reading it." In the following chapters, we continue to leverage our newly developed computing toolbelt to explore more advanced topics of physical interest by continuing to invoke a "reductionist modeler's" approach as in previous chapters.

9.14 Mini-projects

1. Run the tracer1d.py code with $N = 64$ grid cells, and again with $N = 32$ grid cells. What differences do you notice as you decrease the resolution? Explain what you see in terms of the concepts from this chapter.

2. Run the tracer2d.py code with 128×128 grid cells and 64×64 grid cells. Explain what you see in terms of the concepts from this chapter. What impacts are there to the two-dimensional physics that do not have an effect in the one-dimensional case?

3. *"Just-in-Time" compiled code:* The `tracer2d.py` code actually contains two ways to compute the numerical flux. One using a normal scripted approach with `numpy`, the other uses the `numba` library's `@njit()` decorator to invoke a just-in-time compiler to translate the python function to machine code, that is typically hard for humans to read or understand.

   Run the `tracer2d.py` script with 512×512 grid cells and make a note of the simulation's duration (wall clock time) that is printed to the console at the end of the simulation run. Modify the code to call the `upwind_flux_1d_numba` function instead of the `upwind_flux_1d_numpy` function. Which approach took the longest? Repeat this experiment for 1024×1024 grid cells. Which type of computation would you prefer for high-resolution simulations? Explain why.

   What is different about the way the code is written for the two different numerical flux function implementation? Explain why you think the Numba approach is written the way it is. Hint: you can look at the Numba official documentation online for help answering this question.

4. *The Lax-Friedrichs flux:* The Lax-Friedrichs flux is derived by taking a rather blunt approach to introducing numerical smoothing/diffusion to a problem. Consider the 1D tracer equation with a central flux $F_{j+1/2} = \frac{1}{2}\left(cT_j + cT_{j+1}\right)$. The finite volume method's time evolution formula then reads:

$$T_j^{n+1} = T_j^n - \frac{\Delta t}{\Delta x}\left[F_{j+1/2}^n - F_{j-1/2}^n\right].$$

Assuming that there is some numerical noise of wavelength $2\Delta x$ present in the solution at time-level n, one might be tempted to "smear it out" with some averaging, i.e., let $T_j^n = \frac{1}{2}\left(T_{j-1}^n + T_{j+1}^n\right)$ in the above time evolution formula. Substitute this expression into the formula above, and manipulate the equation with algebra so it takes the following form:

$$T_j^{n+1} = T_j^n - \frac{\Delta t}{\Delta x}\left[F_{j+1/2}^{n,*} - F_{j-1/2}^{n,*}\right].$$

What is the formula for $F_{j+1/2}^*$? This new numerical flux function is called the *Lax-Friedrichs flux.* How is the Lax-Friedrichs flux different from the upwind flux for the 1D tracer equation? Back up your answer with an algebraic argument. Will your result still be true for other transport equations, or in two-dimensions? Discuss.

References

[1] D. Alic, et al., Efficient implementation of finite volume methods in numerical relativity, Physical Review D 76 (2007), https://doi.org/10.1103/physrevd.76.104007.

186 CHAPTER 9 Finite volume methods

[2] A. Arakawa, V.R. Lamb, Computational design of the basic dynamical processes of the UCLA general circulation model, Methods in Computational Physics 17 (1977) 173–265.

[3] M. Berger, Chapter 1 – Cut cells: meshes and solvers, in: Remi Abgrall, Chi-Wang Shu (Eds.), Handbook of Numerical Methods for Hyperbolic Problems, in: Handbook of Numerical Analysis, vol. 18, Elsevier, 2017, pp. 1–22, https://www.sciencedirect.com/science/article/pii/S1570865916300394.

[4] P. Brandt, et al., Internal waves in the Strait of Messina studied by a numerical model and synthetic aperture radar images from ERS1/2 satellites, Journal of Physical Oceanography 27 (1997) 648–663.

[5] T.J. Coakley, Implicit Upwind Methods for the Compressible Navier-Stokes Equations, Technical Memorandum 84365, NASA, 1983.

[6] B. Cockburn, C.W. Shu, TVB Runge-Kutta local projection discontinuous Galerkin finite-element method for conservation-laws. 2. General framework, Mathematics of Computation 52 (186) (1989) 411–435, https://doi.org/10.2307/2008474.

[7] A. de la Fuente, et al., The evolution of internal waves in a rotating, stratified, circular basin and the influence of weakly nonlinear and nonhydrostatic accelerations, Limnology and Oceanography 53 (6) (2008) 2738–2748.

[8] D.R. Durran, Numerical Methods for Fluid Dynamics. With Application to Geophysics, 2nd ed., Springer, 2010.

[9] C. Geuzaine, J.-F. Remacle, A three-dimensional finite element mesh generator with built-in pre- and post-processing facilities, International Journal for Numerical Methods in Engineering 79 (2009) 1309–1331.

[10] J. Hesthaven, T. Warburton, Nodal Discontinuous Galerkin Methods, Springer, 2008.

[11] T.J. Hughes, Recent progress in the development and understanding of SUPG methods with special reference to the compressible Euler and Navier-Stokes equations, International Journal for Numerical Methods in Fluids 7 (1987) 1261–1275, https://doi.org/10.1002/fld.1650071108.

[12] R.J. Leveque, Finite Volume Methods for Hyperbolic Problems, Cambridge University Press, 2002.

[13] Randall J. LeVeque, David L. George, High-resolution finite volume methods for the shallow water equations with bathymetry and dry states, in: Advanced Numerical Models for Simulating Tsunami Waves and Runup, 2008, pp. 43–73, https://www.worldscientific.com/doi/abs/10.1142/9789812790910_0002.

[14] S. Vitousek, O.B. Fringer, Physical vs. numerical dispersion in nonhydrostatic ocean modeling, Ocean Modelling 40 (2011) 72–86.

CHAPTER

Modeling fluid transport processes with discontinuous Galerkin methods: Background

10

CONTENTS

10.1	An overview of material	187
10.2	Discontinuous Galerkin finite element method for dispersive shallow water equations	188
10.3	Evaluating the inner products: modes and nodes	192
10.4	Polynomial interpolation nodes in 2D	194
10.5	Local operators for the nodal approach	195
10.6	Surface integral contributions	196
10.7	Boundary conditions	197
10.8	Dealing with source terms: bathymetry and wave dispersion	197
10.9	Solving for the non-hydrostatic pressure: an elliptic problem	200
10.10	Hyperbolic theory in 1D	202
10.11	Linear Riemann problem	203
10.12	Time-stepping method	205
10.13	Mini-projects	206
References		207

10.1 An overview of material

This chapter covers considerable ground in numerical methods. For the reader interested in using the numerical codes, and perhaps less interested in the nuts and bolts of Discontinuous Galerkin Finite Element (DG-FEM) Methods an overview of key points is provided.

1. The basic discretization philosophy is contained in Eqs. (10.10)–(10.13) and the corresponding text.
2. The basic building block used to discretize the geometry and the points at which the solution is given is shown in Fig. 10.1.
3. The unique manner in which our model incorporates nonhydrostatic effects is discussed in Section 10.9.

As alluded to in the closing paragraphs of the previous chapter, general coastal boundaries of bodies of water specify a physical domain with complex/irregular shapes.

Physics and Ecology in Fluids. https://doi.org/10.1016/B978-0-32-391244-0.00020-6
Copyright © 2023 Elsevier Inc. All rights reserved.

188 CHAPTER 10 Discontinuous Galerkin methods: Background

Moreover, if wave dispersion is to be modeled, a rapidly converging method that can resolve small-scale flow features with a minimal set of discretization points is desirable. In this chapter and the one that follows it, we employ the discontinuous Galerkin finite element method (DG-FEM) as a mid-order numerical method to address these concerns. The DG-FEM shares traits with both traditional finite element and finite volume methods that are both commonly used for irregular geometries. The results presented in this chapter use local polynomial orders between $N = 4$ and $N = 8$. The methods are thus "high-order" in contrast to traditional finite element methods that typically use piece-wise linear or quadratic ($N = 2$) basis functions, or the Godunov-type finite volume method that relies on a piece-wise constant representation of the solution on each grid cell. The DG-FEM, however, is "low-order" compared to a pseudo-spectral method [13]. In a nut-shell, by allowing for weak discontinuities at the numerical element interfaces, we gain the flexibility to impose an upwind-biased treatment for the advection terms. The specification of an upwind-biased numerical flux is usually furnished through the well-established theory of approximate Riemann solvers [14] that are commonly used in the formulation of finite volume methods [9] in order to propagate information between finite control volumes. It is for this reason that DG-FEM with piece-wise constant basis functions (order $N = 0$) is identical to the low-order finite volume method [8]. For the practically minded reader, all we are really saying is that we are tapping into a large, well-established literature and picking out a technique that is recent enough to give better results, but not so cutting edge so as to invite controversy.

For the more detail oriented reader, in this chapter, we follow the techniques and developments for nodal DG-FEM as presented by Hesthaven and Warburton [8], building upon their techniques where necessary. In the following sections, we explain the basic nodal DG-FEM formulation as the spatial discretization method for both hyperbolic systems with and without source terms (6.1)–(6.3). We refer the more results-oriented reader to Chapter 11, which provides a discussion of practical issues of how to represent complex boundaries (i.e. the need for a curvilinear treatment of the boundary), as well as an application to the dynamics in a real-world lake, Pinehurst Lake, Alberta, Canada.

10.2 Discontinuous Galerkin finite element method for dispersive shallow water equations

Since the DG-FEM method is primarily suited to solving hyperbolic conservation laws (just like the finite volume method), it is useful to consider the augmented system (6.1)–(6.3) in the conservation form

$$\frac{\partial \vec{Q}}{\partial t} + \frac{\partial \vec{F}}{\partial x} + \frac{\partial \vec{G}}{\partial y} = \vec{B} + \vec{C} + \vec{N}, \tag{10.1}$$

with

$$\vec{Q} = \begin{pmatrix} h \\ hu \\ hv \end{pmatrix}, \quad \vec{F} = \begin{pmatrix} hu \\ hu^2 + \frac{1}{2}gh^2 \\ huv \end{pmatrix}, \quad \vec{G} = \begin{pmatrix} hv \\ huv \\ hv^2 + \frac{1}{2}gh^2 \end{pmatrix}. \qquad (10.2)$$

The terms

$$\vec{B} = gh \begin{pmatrix} 0 \\ \frac{\partial H}{\partial x} \\ \frac{\partial H}{\partial y} \end{pmatrix}, \quad \vec{C} = f \begin{pmatrix} 0 \\ vh \\ -uh \end{pmatrix}, \quad \vec{N} = \frac{H^2}{6} \begin{pmatrix} 0 \\ \nabla \cdot (h\vec{u})_{xt} \\ \nabla \cdot (h\vec{u})_{yt} \end{pmatrix} \qquad (10.3)$$

are the bed slope, the Coriolis pseudo-force, and the wave dispersion terms, respectively.

To apply the DG-FEM method in two dimensions, we first assume that the domain Ω can be triangulated using K elements (or control volumes). The boundary can be approximated in a piece-wise curvilinear sense by using triangles with curved edges as explored later in Section 11.3. Additionally, we assume that the nodes along a triangle edge that are shared between two elements are duplicated, so as to ensure that a purely local scheme can be recovered. This is a fundamental difference between DG-FEM and (continuous) FEM, which uses shared nodes along a shared edge as a means of enforcing strong continuity, and represents a slight trade-off in terms of potential computer memory efficiency.

In each element $\mathbf{D}^k$, we form the approximate local solution $(h^k, (hu)^k, (hv)^k)$ with nodal representations

$$h^k(\vec{x}, t) = \sum_{i=1}^{N_p} h^k(\vec{x}_i^k, t)\ell_i^k(\vec{x}), \qquad (10.4)$$

and similarly for the other fields, hu and hv. Here, $\ell_i^k(x)$ represents the ith order two-dimensional Lagrange interpolating polynomial, $\vec{x} = (x, y)$, and N_p is the number of points within an element. Recall that the 1D Lagrange basis functions are given by

$$\ell_j(x) := \prod_{\substack{0 \le m \le N_p-1 \\ m \ne j}} \frac{x - x_m}{x_j - x_m}$$

$$= \frac{(x - x_0)}{(x_j - x_0)} \cdots \frac{(x - x_{j-1})}{(x_j - x_{j-1})} \frac{(x - x_{j+1})}{(x_j - x_{j+1})} \cdots \frac{(x - x_{N_p-1})}{(x_j - x_{N_p-1})}, \qquad (10.5)$$

and provide a simple and direct way of constructing a polynomial that passes through the points (x_j, y_j) for $j = 0, 1, \cdots, N_p - 1$. The two-dimensional Lagrange basis functions are in general, not known in closed-form and are dependent on the geometry of the domain.

CHAPTER 10 Discontinuous Galerkin methods: Background

For practical reasons, it is assumed that the number of points per element, N_p, is the same for all elements in the domain, although this is not required in theory. The $\mathbf{x}_i^k$'s refer to the local grid points on element $\mathbf{D}^k$ with a distribution that we leave unspecified for the time being.

The nodal DG-FEM weak integral statement is obtained by substituting the approximate local solutions into each of Eqs. (10.1), multiplying by a member of the space of local test functions $V^k = \{\ell_j^k\}_{j=1}^{N_p}$, and integrating the flux terms by parts. If we neglect the $\vec{B}$ and $\vec{N}$ terms in (10.1) for the moment, this gives

$$\int_{\mathbf{D}^k} \frac{\partial \vec{Q}^k}{\partial t} \ell_j^k - \vec{F}^k \frac{\partial \ell_j^k}{\partial x} - \vec{G}^k \frac{\partial \ell_j^k}{\partial y} - \vec{C}^k \ell_j^k dx = -\int_{\partial \tilde{D}^k} \ell_j^k \left(\vec{F}^*, \vec{G}^* \right) \cdot \hat{n} dx \quad (10.6)$$

where $\hat{n}$ is the unit outward-pointing normal. If we were to restrict ourselves to $\ell_0^k = 1$, notice that this formula is simply a more geometrically general version of (9.31) from the previous chapter, in the sense that it does not assume there is a Cartesian ordering to the elements. It is for this reason, that such an approach is called an "unstructured grid" method.

Due to the fact that we do not require the solution to be continuous between elements, the value of $(\vec{F}, \vec{G})$ in the surface integral term on the right-hand side is not unique. Therefore, we have introduced $(\vec{F}^*, \vec{G}^*)$ as the numerical flux vector (see Chapter 9 for further details) that represents some linear combination of information interior to the element $(\vec{F}^-, \vec{G}^-)$ and exterior information $(\vec{F}^+, \vec{G}^+)$. Since we have not explicitly imposed continuity at element interfaces, the numerical flux is our means for imposing continuity in a weak sense. Without it, the elements would completely decouple and a meaningful "global" solution would not be recovered. The numerical flux is typically chosen in a way that "mimics the flow of information in the underlying PDE" to ensure a stable and accurate scheme [8]. The choice of numerical flux mainly considered in this chapter is the local Lax-Friedrichs (L-F) flux (see Mini-Project 8-4, for a 1D derivation)

$$\left(\hat{n}_x \vec{F} + \hat{n}_y \vec{G} \right)^* = \hat{n}_x \{\!\{\vec{F}\}\!\} + \hat{n}_y \{\!\{\vec{G}\}\!\} + \frac{\lambda}{2} [\![\vec{Q}]\!], \quad (10.7)$$

where

$$\{\!\{\vec{u}\}\!\} = \frac{\vec{u}^- + \vec{u}^+}{2}, \qquad [\![\vec{u}]\!] = \hat{n}^- \cdot \vec{u}^- + \hat{n}^+ \cdot \vec{u}^+, \quad (10.8)$$

are the average and jump in an arbitrary field of interest $\vec{u}$ across the interface, respectively. The numerical flux choice (10.7) represents the local Lax-Friedrichs flux where λ is an approximation to the maximum linearized wave speed

$$\lambda = \max_{\vec{s} \in \left[\vec{Q}^-, \vec{Q}^+ \right]} \left(\|\vec{u}(\vec{s})\| + \sqrt{gh(\vec{s})} \right). \quad (10.9)$$

A sketch of the justification for using the Lax-Friedrichs flux, based on the linearized "Riemann problem," is given in Section 10.10.

The "Galerkin approach" outlined above is motivated by attempting to make the residual, i.e., the function left over when the approximate solution is substituted into the PDE system, vanish in a general sense by imposing that the residual should be orthogonal to a space of test functions [8]. Methods that do not take the space of test functions to be the same as the set of basis functions are possible, but are referred to as the more general class of *weighted residual* methods [10]. For example, another possible choice of test functions may be the space of shifted Dirac delta functions centered about a discrete set of points, $\delta(x - x^j)$. The resulting method for the Dirac test functions is known as the *collocation method* that requires the residual to identically vanish at the "collocation points" [10].

In order to reduce the statement (10.6) to a form that is useful for numerical computations, it is important to rewrite it in terms of matrices wherever possible. As an illustration, consider the first component of (10.6) with nodal expansions of the form (10.4) explicitly substituted in. We can then write this first component as

$$\mathcal{M}^k \frac{d\mathbf{h}^k}{dt} = -\mathcal{S}_x^{\mathsf{T},k}(\mathbf{hu})^k - \mathcal{S}_y^{\mathsf{T},k}(\mathbf{hv})^k - \int_{\partial \mathbf{D}^k} \ell_j^k \left((hu)^*, (hv)^*\right) \cdot \hat{\mathbf{n}}d\mathbf{x} \qquad (10.10)$$

where

$$\mathbf{h}^k = \left[h^k(\mathbf{x}_1) \cdots h^k(\mathbf{x}_{N_p})\right]^{\mathsf{T}}, \qquad (10.11)$$

$$(\mathbf{hu})^k = \left[(hu)^k(\mathbf{x}_1) \cdots (hu)^k(\mathbf{x}_{N_p})\right]^{\mathsf{T}}, \qquad (10.12)$$

$$(\mathbf{hv})^k = \left[(hv)^k(\mathbf{x}_1) \cdots (hv)^k(\mathbf{x}_{N_p})\right]^{\mathsf{T}}, \qquad (10.13)$$

and we have left the surface integral contribution alone for now. Here the local mass matrix is given by

$$\mathcal{M}_{ij}^k = \int_{\mathbf{D}^k} \ell_i^k(\mathbf{x})\ell_j^k(\mathbf{x})d\mathbf{x} = J^k \int_{\mathbf{I}} \ell_i(\mathbf{r})\ell_j(\mathbf{r})d\mathbf{r} = J^k \mathcal{M}, \qquad (10.14)$$

where $J^k = x_r^k y_s^k - x_s^k y_r^k$ is the (constant) Jacobian of the linear mapping from the element $\mathbf{D}^k$ to the reference right triangular element $\mathbf{I} = \{\mathbf{r} = (r, s) | (r, s) \geq -1; r + s \leq 0\}$, and we have also introduced the mass matrix on the reference right triangle, $\mathcal{M}$.

The local stiffness matrix $\mathcal{S}_x^k$ is given as follows:

$$\mathcal{S}_{x,ij}^k = \int_{\mathbf{D}^k} \ell_i^k(\mathbf{x})\frac{\partial \ell_j^k}{\partial x}d\mathbf{x} = J^k \int_{\mathbf{I}} \ell_i(\mathbf{r}) \left(\frac{\partial \ell}{\partial r}r_x^k + \frac{\partial \ell}{\partial s}s_x^k\right)d\mathbf{r}, \qquad (10.15)$$

$$= J^k \int_{\mathbf{I}} \ell_i(\mathbf{r}) \left(\frac{\partial \ell}{\partial r}\frac{y_s^k}{J^k} - \frac{\partial \ell}{\partial s}\frac{y_r^k}{J^k}\right)d\mathbf{r}, \qquad (10.16)$$

$$= y_s^k \mathcal{S}_r - y_r^k \mathcal{S}_s, \qquad (10.17)$$

192 **CHAPTER 10** Discontinuous Galerkin methods: Background

where we have used the fact that the Jacobian matrices have the inverse property, i.e.,

$$\frac{\partial \mathbf{x}}{\partial \mathbf{r}}\frac{\partial \mathbf{r}}{\partial \mathbf{x}} = \begin{bmatrix} x_r & x_s \\ y_r & y_s \end{bmatrix}\begin{bmatrix} r_x & r_y \\ s_x & s_y \end{bmatrix} = \begin{bmatrix} 1 & 0 \\ 0 & 1 \end{bmatrix} , \tag{10.18}$$

hence,

$$r_x = \frac{y_s}{J} , \ r_y = -\frac{x_s}{J} , \ s_x = -\frac{y_r}{J} , \ s_y = \frac{x_r}{J} , \tag{10.19}$$

in going from (10.15) to (10.16). Similarly, for $\mathcal{S}_y^k$, we have

$$\mathcal{S}_{y,ij}^k = \int_{\mathbf{D}^k} \ell_i^k(\mathbf{x})\frac{\partial \ell_j^k}{\partial x}d\mathbf{x} = -x_s^k \mathcal{S}_r + x_r^k \mathcal{S}_s . \tag{10.20}$$

The stiffness matrices defined on the standard triangle $\mathbf{I}$ are given by

$$\mathcal{S}_{r,ij} = \int_{\mathbf{I}} \ell_i(\mathbf{r})\frac{\partial \ell_j}{\partial r}d\mathbf{r} , \quad \mathcal{S}_{s,ij} = \int_{\mathbf{I}} \ell_i(\mathbf{r})\frac{\partial \ell_j}{\partial s}d\mathbf{r} . \tag{10.21}$$

We have hence written all local mass and stiffness matrices in terms of inner products over the standard triangle $\mathbf{I}$. For the moment, however, it is unclear how to evaluate these inner products since the explicit form of the two-dimensional Lagrange polynomials on a triangle is not known [8]. The developments by Hesthaven and Warburton [8] ensure that the evaluation of these inner products can be performed implicitly by considering an expansion in appropriately-chosen basis function that can be evaluated in a general way for arbitrary orders of approximation.

10.3 Evaluating the inner products: modes and nodes

We follow the discussion in [8] by introducing a modal basis-function expansion for each solution field as an alternative to the nodal representations (e.g. (10.4)). For example, for an arbitrary approximate field $u(\mathbf{r})$ defined on $\mathbf{I}$, we have

$$u(\mathbf{r}) = \sum_{n=1}^{N_p} \hat{u}_n \psi_n(\mathbf{r}) = \sum_{i=1}^{N_p} u(\mathbf{r}_i)\ell_i(\mathbf{r}) , \tag{10.22}$$

where $\{\psi(\mathbf{r})\}_{i=1}^{N_p}$ is a two-dimensional basis. The relationship between the modes $\hat{u}_n$ and the nodes $u(\mathbf{r}_i)$ can be established by projecting onto a particular member of the basis ψ_m, i.e.,

$$\int_{\mathbf{I}} u(\mathbf{r})\psi_m(\mathbf{r})d\mathbf{r} = \sum_{n=1}^{N_p} \hat{u}_n \int_{\mathbf{I}} \psi_n(\mathbf{r})\psi_m(\mathbf{r})d\mathbf{r} , \tag{10.23}$$

10.3 Evaluating the inner products: modes and nodes   **193**

or, in matrix-vector notation,

$$\mathbf{v} = \mathcal{H}\hat{\mathbf{u}} , \tag{10.24}$$

where

$$\hat{\mathbf{u}} = [\hat{u}_1, \cdots , \hat{u}_{N_p}], \quad \mathcal{H}_{ij} = \int_{\mathbf{I}} \psi_i \psi_j d\mathbf{r}, \quad \mathbf{v}_i = \int_{\mathbf{I}} u \psi_i d\mathbf{r} . \tag{10.25}$$

In order to ensure that $\mathcal{H}$ is well-conditioned (i.e., the basis functions are well-behaved) for an arbitrary-sized basis, it is instructive to choose the basis $\{\psi(\mathbf{r})\}_{i=1}^{N_p}$ to be orthonormal. That is, dot-products with one another should be zero, and dot-products with themselves should be 1. It is then clear that $\mathcal{H}$ will reduce to the identity matrix. An appropriate basis can be found by applying the Gram-Schmidt process from linear algebra to the simpler monomial basis $r^i s^j$ where $0 \le i + j \le N$. The details of this re-normalization process are rather involved, and we simply quote the result stated in [8] here, as

$$\psi_m(\mathbf{r}) = \sqrt{2} P_i(a) P_j^{(2i+1,0)}(b)(1-b)^i , \tag{10.26}$$

where

$$a = 2\frac{1+r}{1-s} - 1, \quad b = s , \tag{10.27}$$

and $P_n^{(\alpha,\beta)}$ is the nth-order Jacobi polynomial. As a special case, the polynomial with $\alpha = \beta = 0$, $P_n = P_n^{(0,0)}$, is the nth-order Legendre polynomial. In one space dimension, the relationship between the order of the highest-degree basis polynomial and the number of points on the element is given by $N_p = N + 1$. On the triangle, however, the relationship is given by the $(N + 1)^{st}$ triangular number

$$N_p = \sum_{i=1}^{N+1} i = \binom{N+2}{2} , \tag{10.28}$$

that can be derived by counting the number of basis polynomials of degree at most N.

The only remaining question is how to evaluate the inner (or dot-products) on the left-hand side of the projection (10.23). In the nodal approach [8], we assume the modal expansion interpolates u at the nodes $\mathbf{r}_i$, i.e.,

$$u(\mathbf{r}_i) = \sum_{n=1}^{N_p} \hat{u}_n \psi_n(\mathbf{r}_i) . \tag{10.29}$$

It follows that the relationship between "the nodes" and "the modes" can be established via the generalized Vandermonde matrix $\mathcal{V}$, that is

$$\mathcal{V}\hat{\mathbf{u}} = \mathbf{u} , \tag{10.30}$$

where $\mathcal{V}_{ij} = \psi_j(\mathbf{r}_i)$, $\hat{\mathbf{u}}_i = \hat{u}_i$, and $\mathbf{u}_i = u(\mathbf{r}_i)$. The standard (i.e., not "generalized") Vandermonde matrix supplies a mapping between the monomial modal coefficients to some polynomial function value. The word "generalized" appears here, since the concept has been applied instead to orthonormal polynomials.

Combining (10.30) with the uniqueness statement (10.22), one can obtain the following formula for the Lagrange polynomials in terms of the basis polynomials

$$\ell_i(\mathbf{r}) = \sum_{n=1}^{N_p} \left(\mathcal{V}^{\mathsf{T}}\right)^{-1}_{in} \psi_n(\mathbf{r}) . \qquad (10.31)$$

While this may feel to the reader to be a "Eureka" moment, Eq. (10.31) is hardly useful in practice. Rather, it simply represents a combination of our well-behaved (or orthonormal) basis functions with our simplifying assumptions about our interpolation points. This expression is only useful for deriving expressions for the integrals and derivatives in the rather "theory-heavy" DG-FEM formulation.

10.4 Polynomial interpolation nodes in 2D

It remains to ask what a proper choice of points $\mathbf{r}_i$ should be to ensure that the system (10.30) is well-conditioned for arbitrary orders of approximation N. What we are in effect demanding is that our method "always works" (or is at least consistent with the continuous problem, and converges to its solution, eventually), no matter what we choose for $N > 0$.

In one dimension, one can show [8] that the "optimal" choice of points for the orthogonal Legendre polynomials is the Legendre-Gauss-Lobotto (LGL). The LGL nodes can be computed quickly by numerically calculating the eigenvalues of a tridiagonal system [8,15] and they ensure good conditioning by requiring the determinant of $\mathcal{V}$ to grow unbounded monotonically for large N.

On the triangle, the optimal choice of points is much more difficult to derive analytically. Until recently, this represented an open problem with the calculation of nodal sets on the triangle often leading to expensive implicit computations. The method of Warburton [16] considers a simple and inexpensive explicit approach for selecting the points, and we briefly explain this technique here. The resulting points are referred to as "near-optimal" since optimality has not been theoretically proven. However, the desired growth of the Vandermonde matrix determinant has been established [16]. The resulting set of points are not associated with quadrature formulas like the 1D LGL quadrature nodes, but are closely related to the one-dimensional LGL nodes. An illustration of these points on an equilateral triangle for polynomial order $N = 12$ is shown in Fig. 10.1.

The underlying idea behind choosing the interpolation points on the triangle is to use the fact that along a triangle's edge, we expect the nodes to be distributed by arc-length in the same manner as the 1D LGL nodes on the interval $[-1, 1]$. Using

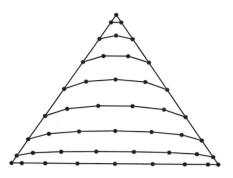

FIGURE 10.1

Warburton's "near-optimal" polynomial interpolation nodes on an equilateral triangle for order $N = 8$.

this fact, a 1D deformation field called the *warp factor* can be computed that describes how the 1D equidistant nodes should be displaced to arrive at the LGL nodes along the edge. The final step relies on a technique known as Gordon-Hall blending. This technique blends the 1D deformation to the interior points by choosing an appropriate weighting function (called the *blend factor*) in the edge normal direction. The blend factor is based on barycentric coordinates to obey symmetry properties of the equilateral triangle [7,8]. The resultant 2D displacement field is then applied to the equidistant points on the triangle, and the process is repeated for all edges of the triangle.

Though the method of Gordon and Hall was originally intended for constructing curvilinear coordinate systems, and we use it for this very purpose later on in Section 11.3, it has been shown to be useful as well for the purpose of choosing polynomial interpolation nodes in 2D and 3D geometries, and in particular on the triangle [8,16].

10.5 Local operators for the nodal approach

We can now apply all of the developments of the above section to the evaluation of the inner products posed by Eqs. (10.14) and (10.21). Substituting the formula (10.31) into the expression for the standard local mass matrix, we recover

$$\mathcal{M} = \left(\mathcal{V}\mathcal{V}^\mathsf{T}\right)^{-1}. \tag{10.32}$$

Before considering the stiffness matrices, it is useful to define the differentiation matrices

$$\mathcal{D}_{r,ij} = \left.\frac{\partial \ell_j}{\partial r}\right|_{\mathbf{r}_i}, \quad \mathcal{D}_{s,ij} = \left.\frac{\partial \ell_j}{\partial s}\right|_{\mathbf{r}_i} \tag{10.33}$$

196   **CHAPTER 10** Discontinuous Galerkin methods: Background

whose entries may be furnished directly by appropriate differentiation of the formula (10.31). It can be shown [8] that the local stiffness matrices can be recovered by

$$\mathcal{M}\mathcal{D}_r = \mathcal{S}_r, \quad \mathcal{M}\mathcal{D}_s = \mathcal{S}_s . \tag{10.34}$$

In other words,

$$\mathcal{D}_r = \mathcal{M}^{-1}\mathcal{S}_r, \quad \mathcal{D}_s = \mathcal{M}^{-1}\mathcal{S}_s . \tag{10.35}$$

This result is useful because it implies that an explicit semi-discrete scheme can be obtained by multiplying (10.10) by $(\mathcal{M}^k)^{-1} = \frac{1}{J^k}\mathcal{M}^{-1}$. As a consequence of the fact that the local mass matrix only varies by a constant factor on each element, it follows that this operation is computationally cheap since $\mathcal{M}$ is an $N_p \times N_p$ matrix. For example, with order $N = 8$ basis functions, the local mass matrix is a 45×45 full matrix. This is another key difference between DG-FEM and the classical FEM, where explicit semi-discrete schemes often cannot be recovered since the time-derivative operator is multiplied by the global mass matrix, that may be large and expensive to invert explicitly. It is worth mentioning that if inexact quadrature rules are applied to evaluate the integrals, the global mass matrix is diagonalized in the classical FEM [6], and the issue is avoided.

10.6 Surface integral contributions

To close our numerical scheme, it remains to discuss the surface integral term in Eq. (10.10)

$$-\int_{\partial \mathbf{D}^k} \ell_j^k(\mathbf{x})\mathbf{g} \cdot \hat{\mathbf{n}}d\mathbf{x} \tag{10.36}$$

where $\mathbf{g} = ((hu)^*, (hv)^*)$ represents the flux across an interface. Since the normal $\hat{\mathbf{n}}$ is constant along each edge, it is useful to break this expression up into three integrals

$$-\int_{\partial \mathbf{D}^k} \ell_j^k(\mathbf{x})\mathbf{g} \cdot \hat{\mathbf{n}}d\mathbf{x} = -\sum_{e=1}^{3} \hat{\mathbf{n}}_e \cdot \int_{\text{edge}_e} \ell_j^k(\mathbf{x})\mathbf{g}d\mathbf{x} . \tag{10.37}$$

If we substitute the nodal expression $\mathbf{g} = \sum_{i=1}^{N+1} \ell_i^k(\mathbf{x})\mathbf{g}_i$ the right hand side reduces to

$$-\sum_{e=1}^{3} \sum_{i=1}^{N+1} \hat{\mathbf{n}} \cdot \mathbf{g}_i \mathcal{M}_{ij}^{k,e} , \tag{10.38}$$

where we have introduced the $(N + 1) \times (N + 1)$ edge mass matrix

$$\mathcal{M}_{ij}^{k,e} = \int_{\text{edge}_e} \ell_j^k(\mathbf{x})\ell_i^k(\mathbf{x}) \, d\mathbf{x} = J^{k,e,1}\mathcal{M}^1 , \tag{10.39}$$

where $J^{k,e,1}$ is the Jacobian of the mapping from the edge to the standard interval $[-1, 1]$. Using the 1D developments in [8], the standard 1D mass matrix is related to the Vandermonde matrix for 1D polynomial interpolation by $\mathcal{M}^1 = \left(\mathcal{V}^1 (\mathcal{V}^1)^T\right)^{-1}$.

10.7 Boundary conditions

The freedom in the numerical flux choice gives us a convenient way to impose boundary conditions through appropriately choosing imaginary "ghost" states (see Chapter 9 for "ghosting" in the finite volume context), i.e. the "+" traces along boundary edges. For a purely reflective wall with no flow going through it, we impose

$$h^+ = h^- , \tag{10.40}$$

$$hu^+ = hu^- - 2(n_x hu + n_y hv)n_x , \tag{10.41}$$

$$hv^+ = hv^- - 2(n_x hu + n_y hv)n_y . \tag{10.42}$$

The second and third conditions are equivalent to imposing no normal $\mathbf{u} \cdot \hat{\mathbf{n}} = 0$ along the wall. The first condition is equivalent to imposing $\nabla h \cdot \hat{\mathbf{n}} = 0$ which states that the interface should be parallel to the bathymetry at the wall.

10.8 Dealing with source terms: bathymetry and wave dispersion

So far, we have not discussed the treatment of the bathymetry and nonhydrostatic (wave dispersion) terms contained in the vectors $\mathbf{B}$ and $\mathbf{N}$, respectively. We have avoided these terms so far because they cannot be addressed by the standard nodal DG-FEM treatment.

As an example of the issues that arise, let us consider the second entry of $\mathbf{B}$. If we multiply by ℓ_j^k, and integrate over the element, the following integrals appear in the weak DG statement

$$\int_{\mathbf{D}^k} gh^k \frac{\partial H^k}{\partial x} \ell_j(x) d\mathbf{x} - \int_{\partial \mathbf{D}^k} gh H^* d\mathbf{x} , \tag{10.43}$$

the surface integral term does not pose a problem, and in the case where H is continuous across element interfaces, it vanishes. The first term, on the other hand does pose a problem because we cannot write it in terms of the local stiffness matrix $\mathcal{S}_x^k$. To see this, let us substitute a nodal expansion in for H, yielding (after index relabeling)

$$\int_{\mathbf{D}^k} gh^k \frac{\partial H^k}{\partial x} \ell_j(\mathbf{x}) d\mathbf{x} = \sum_{j=1}^{N_p} g H^k(x_j) \int_{\mathbf{D}^k} h^k \ell_i(\mathbf{x}) \frac{\partial \ell_j}{\partial x} d\mathbf{x} ,$$

$$= \sum_{j=1}^{N_p} g H^k(x_j) \mathcal{S}_{ij}^{\mathsf{T},k,h}, \tag{10.44}$$

$$= g \mathcal{S}^{\mathsf{T},k,h} \mathbf{H}^k, \tag{10.45}$$

where we have taken the integral on the right to be the modified local stiffness matrix, which depends on the water depth, h. Since h is a function of both space and time, this approach is computationally more expensive since the local stiffness matrix is now different on every element and must be updated after each time-step. Although this approach is more computationally expensive in general, it should not be completely disregarded since it becomes necessary in situations where curvilinear elements are used since the mapping Jacobian is no longer constant. See Sections 11.3–11.5 for further explanation.

To pursue a less expensive approach, let us introduce the auxiliary variable

$$\kappa(\mathbf{x}) = \frac{\partial H}{\partial x}. \tag{10.46}$$

Following previous discussion, we can approximate κ with the DG-FEM method by

$$\mathcal{M}^k \boldsymbol{\kappa} = \mathcal{S}_x^{\mathsf{T}} \mathbf{H}^k - \int_{\partial \mathbf{D}^k} H^* n_x \, d\mathbf{x}, \tag{10.47}$$

or,

$$\boldsymbol{\kappa} = \mathcal{D}_x^{\mathsf{T}} \mathbf{H}^k - \left(\mathcal{M}^k\right)^{-1} \int_{\partial \mathbf{D}^k} H^* n_x \, d\mathbf{x}. \tag{10.48}$$

If we now return to the bathymetry terms, we are charged with computing the integral

$$\int_{\mathbf{D}^k} g h^k(\mathbf{x}) \kappa^k(\mathbf{x}) \ell_j(\mathbf{x}) d\mathbf{x}. \tag{10.49}$$

We could proceed as before and simply substitute in the nodal expansion for κ^k, we would then be left with a modified mass matrix $\mathcal{M}^{k,h}$, and we will not have gained much. On the other hand, if we approximate the nodal expansion product $h^k \kappa^k$ in the following manner

$$h^k(\mathbf{x}) \kappa^k(\mathbf{x}) \approx \sum_{i=1}^{N_p} h^k(\mathbf{x}_i) \kappa^k(\mathbf{x}_i) \ell_i^k(\mathbf{x}), \tag{10.50}$$

i.e., we approximate the function product with a *point-wise product*, we then recover the scheme

$$\int_{\mathbf{D}^k} g h^k \frac{\partial H^k}{\partial x} \ell_j(\mathbf{x}) d\mathbf{x} \approx \sum_{j=1}^{N_p} g \kappa^k(\mathbf{x}_j) h^k(\mathbf{x}_j) \int_{\mathbf{D}^k} \ell_i(\mathbf{x}) \ell_j(\mathbf{x}) d\mathbf{x},$$

$$= g \mathcal{M}^k (\boldsymbol{\kappa} \mathbf{h})^k, \tag{10.51}$$

10.8 Dealing with source terms: bathymetry and wave dispersion

which is less computationally expensive than the former scheme since the local mass matrix only varies by a constant value between elements.

The price we pay when using this approximation is that we have essentially committed a couple of "variational crimes" [8]. Aliasing errors result from two distinct sources: 1) the fact that a product of two functions cannot be completely recovered by a point-wise product between the nodal values (see Mini-Project #4); and 2) the fact that the interpolant of a derivative is not (numerically speaking) the same thing as the derivative of an interpolant. Filtering of the modal coefficients can be used to prevent these aliasing errors from driving weak numerical instabilities. The form of the filter we choose is an exponential taken to the power α in the order of the basis polynomial. Here, α is a tunable "filtering parameter" as is the cut-off polynomial order, which we denote as N_c.

The inexpensive nodal approach presented here is used to time-step both the bathymetry terms $gh\nabla H$ and the non-hydrostatic terms under the change of variables $z = \nabla \cdot (h\mathbf{u})_t$ and $\gamma = H^2/6$, such that the term takes a familiar (pressure gradient) form of $\gamma\nabla z$.

Both of these terms may be regarded as source terms in the DG-FEM formulation assuming that h and z are known. The gradient of z (the non-hydrostatic part of the pressure gradient) may be either evaluated using the central flux $z^* = \{\!\{z\}\!\}$ or the purely internal choice $z^* = z^-$. At this point however, the auxiliary scalar field z is not a known quantity to the reader and a governing equation must be derived.

The momentum equations for the 2D dispersive shallow water model are given by

$$\frac{\partial (hu)}{\partial t} + \nabla \cdot \left(hu^2, huv\right)^{\mathsf{T}} - fhv = -gh\eta_x + \frac{H^2}{6}\left(\nabla \cdot (h\vec{u})\right)_{xt}, \qquad (10.52)$$

$$\frac{\partial (hv)}{\partial t} + \nabla \cdot \left(huv, hv^2\right)^{\mathsf{T}} + fhu = -gh\eta_y + \frac{H^2}{6}\left(\nabla \cdot (h\vec{u})\right)_{yt}. \qquad (10.53)$$

Taking the divergence of the momentum equations (10.52)–(10.53) we arrive at the elliptic equation

$$\nabla \cdot (\gamma\nabla z) - z = -\nabla \cdot \vec{a}, \qquad (10.54)$$

that is referred to as a *wave continuity* equation in [4]. The vector $\vec{a} = (a_1, a_2)^T$ is given by the flux terms in Eqs. (6.2)–(6.3), i.e., in component form the vector reads

$$\vec{a} = \begin{pmatrix} -\nabla \cdot ((hu)\vec{u}) - gh\eta_x + fhv \\ -\nabla \cdot ((hv)\vec{u}) - gh\eta_y - fhu \end{pmatrix}. \qquad (10.55)$$

For the more advanced or diligent reader, we explain how z is calculated by solving the elliptic problem (10.54) within the DG-FEM framework below.

200 CHAPTER 10 Discontinuous Galerkin methods: Background

10.9 Solving for the non-hydrostatic pressure: an elliptic problem

At first, it may not be clear how the DG-FEM can be applied to second-order elliptic equations since such equations are not hyperbolic and thus do not have well-posed Riemann problems that may be considered to impose weak continuity across element interfaces. However, it is explained in [8] that elliptic equations may be recast as a first-order system of equations and appropriate numerical fluxes can be obtained from the class of numerical "penalty methods."

To re-write (10.54) as a first-order system, we introduce the auxiliary variable

$$\vec{q} = (q_x, q_y) = \sqrt{\gamma}\nabla z, \tag{10.56}$$

yielding the system

$$\nabla \cdot \left(\sqrt{\gamma}\vec{q}\right) - z = -\nabla \cdot \vec{a}, \tag{10.57}$$

$$q_x = \sqrt{\gamma}\frac{\partial z}{\partial x}, \tag{10.58}$$

$$q_y = \sqrt{\gamma}\frac{\partial z}{\partial y}. \tag{10.59}$$

Inspecting the system (10.57)–(10.59) it may be unclear how, given an input right-hand side $-\nabla \cdot \vec{a}$, one can recover z. This is generally achieved by considering the inverse situation, i.e., if z is known, then $\vec{q}$ can be computed by solving Eqs. (10.58)–(10.59), and $-\nabla \cdot \vec{a}$ can be recovered using (10.57). This set of operations can be considered a non-singular linear transformation, and hence there must exist an inverse transformation.

The strong DG formulation, where integration by parts is carried out twice, of (10.58)–(10.59) together with the weak formulation of (10.57) is given by

$$\mathcal{M}^k \mathbf{q_x}^k = \sqrt{\boldsymbol{\gamma}}^k \mathcal{S}_x \mathbf{z}^k - \sqrt{\boldsymbol{\gamma}}^k \int_{\partial \mathbf{D}^k} \ell_j^k \left(z^k - z^*\right) n_x \, d\mathbf{x}, \tag{10.60}$$

$$\mathcal{M}^k \mathbf{q_y}^k = \sqrt{\boldsymbol{\gamma}}^k \mathcal{S}_y \mathbf{z}^k - \sqrt{\boldsymbol{\gamma}}^k \int_{\partial \mathbf{D}^k} \ell_j^k \left(z^k - z^*\right) n_y \, d\mathbf{x}, \tag{10.61}$$

$$-(\mathcal{S}_x^k)^\mathsf{T} \left(\sqrt{\boldsymbol{\gamma}}\mathbf{q_x}\right)^k - (\mathcal{S}_y^k)^\mathsf{T} \left(\sqrt{\boldsymbol{\gamma}}\mathbf{q_y}\right)^k + \int_{\partial \mathbf{D}^k} \ell_j^k (\sqrt{\boldsymbol{\gamma}}\mathbf{q})^* \cdot \hat{\mathbf{n}} \, d\mathbf{x} - \mathcal{M}^k \mathbf{z}^k =$$
$$(\mathcal{S}_x^k)^\mathsf{T} \mathbf{a_x}^k + (\mathcal{S}_y^k)^\mathsf{T} \mathbf{a_y}^k - \int_{\partial \mathbf{D}^k} \ell_j^k \mathbf{a}^* \cdot \hat{\mathbf{n}} \, d\mathbf{x}. \tag{10.62}$$

Here, we choose the central flux for the right-hand side, i.e. $\mathbf{a}^* = \{\!\{\mathbf{a}\}\!\}$ together with the interior penalty (IP) flux for the elliptic operator, i.e. $z^* = \{\!\{z\}\!\}$, $(\sqrt{\gamma}\mathbf{q})^* = \{\!\{\sqrt{\gamma}\nabla z\}\!\} - \tau[\![z]\!]$, $\tau > 0$. The point of the penalty term is to penalize large jumps

10.9 Solving for the non-hydrostatic pressure: an elliptic problem

at the element interfaces. If $\tau = 0$, a numerical calculation of the eigenfunctions of the Laplacian would reveal a spurious $\lambda = 0$ mode with all elements completely decoupled, and the system would be singular [8]. The use of the penalty term pushes this spurious eigenfunction out of the operator's null space to guarantee invertibility. In general, a sufficiently large penalty parameter will suppress any other spurious modes to the high-λ part of the eigenspectrum as well. This property represents an advantage over continuous Galerkin discretizations of elliptic operators that often possess spurious *convergent* modes whose corresponding eigenvalues can lie within the physical range of the eigenspectrum [3].

Other flux possibilities for DG-discretized Laplacian and Helmholtz operators include the penalized central flux $z^* = \{\!\{z\}\!\}$, $(\sqrt{\gamma}\mathbf{q})^* = \{\!\{\sqrt{\gamma}\mathbf{q}\}\!\} - \tau [\![z]\!]$ and the local discontinuous Galerkin (LDG) flux $z^* = \{\!\{z\}\!\} + \hat{\mathbf{n}}[\![z]\!]$, $(\sqrt{\gamma}\mathbf{q})^* = \{\!\{\sqrt{\gamma}\mathbf{q}\}\!\} - [\![\sqrt{\gamma}\mathbf{q}]\!] \cdot \hat{\mathbf{n}} - \tau [\![z]\!]$. The central flux should not be used in general since it only shows optimal convergence characteristics for even polynomial orders N, and its matrix-form is not as sparse as the other choices (i.e., large computational stencil). The LDG flux, on the other hand, has optimal convergence rates at all orders and offers the most sparse representation, but it is known to be the most poorly conditioned operator. The IP flux offers a balance between these two, giving optimal convergence at all orders, a middle-ground in terms of sparsity, and similar condition numbers to the central-flux operator [8]. With some algebraic manipulations, the auxiliary variable $\mathbf{q}$ can be eliminated locally, and this allows the operator to be efficiently set-up directly as a symmetric sparse matrix. Such an elimination gives the following statement in terms of local operators

$$
\begin{aligned}
-&\left(\left(\mathcal{D}_x^k \right)^T \mathcal{M}^k \Gamma^k \left(\mathcal{M}^k \right)^{-1} \Gamma^k \mathcal{M}^k \mathcal{D}_x^k \right. \\
&+\left(\mathcal{D}_y^k \right)^T \mathcal{M}^k \Gamma^k \left(\mathcal{M}^k \right)^{-1} \Gamma^k \mathcal{M}^k \mathcal{D}_y^k + \mathcal{M}^k \right) \mathbf{z}^k \\
&+\sum_{e=1}^{3} \left(\mathcal{D}_n^{k,e} \right)^T \mathcal{M}^{k,e} \Gamma^{k,e} \left(\mathcal{M}^{k,e} \right)^{-1} \Gamma^{k,e} \mathcal{M}^{k,e} \left(\frac{\mathbf{z}^- - \mathbf{z}^+}{2} \right) \qquad (10.63) \\
&+\sum_{e=1}^{3} \mathcal{M}^{k,e} \left[\Gamma^{k,e} \left(\mathcal{D}_n^{k,e} \left(\frac{\mathbf{z}^- - \mathbf{z}^+}{2} \right) + \tau (\mathbf{z}^- - \mathbf{z}^+) \right) \right] = \mathrm{RHS} ,
\end{aligned}
$$

where Γ^k is the diagonal matrix with the entries of $\sqrt{\gamma}^k$ written along its diagonal and $\mathcal{D}_n^{k,e} = \mathcal{D}_x^k n_x^{k,e} + \mathcal{D}_y^k n_y^{k,e}$ is the discretized normal derivative along edge e of element k.

The discontinuous Galerkin IP discretization method has become known as the "symmetric interior penalty discontinuous Galerkin" (SIP-DG) method in the literature, and has been applied to the pressure Poisson equation and viscous operator of the incompressible Navier-Stokes equations [5,11].

202 CHAPTER 10 Discontinuous Galerkin methods: Background

10.10 Hyperbolic theory in 1D

Many numerical methods, including the finite volume method of Chapter 9 and discontinuous Galerkin [2] methods, rely on the well-established hyperbolic theory of conservation laws to exchange information between computational cells, since hyperbolic theory gives insights into how information propagates as time evolves. The ideas are based upon the method of characteristics that, in 1D, looks for curves in the x-t plane along which the solution is constant, called *characteristic curves* or simply *characteristics*. Once the characteristic curves are known, the full solution can be determined by following the characteristics forward in time beginning from the initial data.

Although many PDE systems of practical interest are not fully hyperbolic (e.g., dispersive wave equations or equations for incompressible flow), their numerical treatments typically "split" terms up [1,4] such that at least one step in the solution procedure resembles solving a hyperbolic system of equations, and the importance of understanding some hyperbolic theory for the purposes of numerical methods is thus clear. In this section, we give a brief introduction to the utility of hyperbolic theory for the 1D linear shallow water equations.

The 1D linear shallow water equations can be written as

$$\vec{q}_t + \left(\vec{F}(\vec{q}) \right)_x = \vec{S}(\vec{q}) , \tag{10.64}$$

where $\vec{q} = (h, u)^{\mathsf{T}}$, and the flux vector is given by

$$\vec{F}(\vec{q}) = \begin{pmatrix} Hu \\ gh \end{pmatrix} . \tag{10.65}$$

$\vec{S}(\vec{q})$ corresponds to source terms, which in this context would be due to variable bathymetry. Since source terms do not affect the characteristics of the system, for simplicity we can assume a flat bottom, hence $\vec{S}(\vec{q}) = 0$. Applying the chain rule to (10.64) gives

$$\vec{q}_t + A\vec{q}_x = 0 , \tag{10.66}$$

where

$$A = \frac{\partial \vec{F}}{\partial \vec{q}} , \tag{10.67}$$

is the flux Jacobian. Here,

$$A = \begin{pmatrix} 0 & H \\ g & 0 \end{pmatrix} . \tag{10.68}$$

A first-order system is said to be hyperbolic if A is diagonalizable with real eigenvalues. The characteristic wave speeds are then given by the eigenvalues $\lambda_{1,2}$ of A. This fact can be seen by realizing that diagonalizing the PDE system would result in a set of de-coupled linear advection equations each with its own linear wave speed. This result can be further motivated by presuming that a wave speed λ is known and making the coordinate substitution $\xi = x - \lambda t$, in which case the system would read

$$\begin{pmatrix} -\lambda & H \\ g & -\lambda \end{pmatrix} \begin{pmatrix} h \\ u \end{pmatrix}_\xi = 0 \,, \tag{10.69}$$

and it becomes clear that interesting solutions require the determinant of the matrix in Eq. (10.69) to be zero. The characteristic equation $\lambda^2 - gH = 0$ is solved quite simply to give the characteristic wave speeds

$$\lambda_1 = -\sqrt{gH} \,, \quad \lambda_2 = \sqrt{gH} \,, \tag{10.70}$$

corresponding to leftward and rightward propagating waves, respectively. The eigenvectors $\vec{v}_{1,2}$ can also be determined from their definition $A\vec{v} = \lambda\vec{v}$, and since they are determined only up to a multiplicative constant we can choose their first entry be 1 without loss of generality. Hence,

$$\vec{v}_1 = \begin{pmatrix} 1 \\ -\sqrt{g/H} \end{pmatrix} \,, \quad \vec{v}_2 = \begin{pmatrix} 1 \\ \sqrt{g/H} \end{pmatrix} \,. \tag{10.71}$$

From a mathematical point of view, the eigenvectors dictate what the weighting of flow variables should be such that the solution profile is a single translating wave for all times. For example, if $h(x, 0)$ is the initial depth profile, then setting the initial velocity to $u(x, 0) = \left(\sqrt{g/H}\right) h(x, 0)$[1] would result in a solution profile that translates to the right with the linear long wave speed $\lambda_2 = \sqrt{gH}$. From a physics perspective, the eigenvectors dictate what the current induced by a single wave propagating either to the left or right would be.

10.11 Linear Riemann problem

The Riemann problem is a well-studied applied mathematics problem that asks the question of how an initial jump between two constant states would evolve forward in time in a hyperbolic PDE system, and it is fundamental in the formulation of numerical methods where information must be exchanged between computational cells when strong continuity is not enforced at cell interfaces.

Without loss of generality, we may assume that the jump is centered at $x = 0$ and that the initial data is given by

$$\vec{q}(x, 0) = \begin{cases} \vec{q}_l \,, & x \le 0 \,, \\ \vec{q}_r \,, & x > 0 \,. \end{cases} \tag{10.72}$$

---

[1] One could also take $u(x, 0) = \sqrt{g/H}\eta(x, 0)$ so that there is no mean background flow.

As shown in Fig. 10.2, the method of characteristics implies that for the linear shallow water system the solution should split into three regions for $t > 0$, corresponding to the initial left $\vec{q}_l$ and right states $\vec{q}_r$, and an intermediate state $\vec{q}^*$. We are charged with determining the value of $\vec{q}^*$ so that the full solution will be known for all times following the initialization.

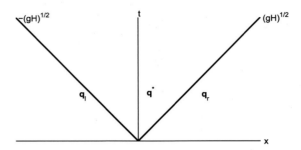

FIGURE 10.2

Illustration of the Riemann problem in the xt-plane. Immediately after the initialization, a third state ($\vec{q}^*$) appears in the solution that must be determined.

As explained in [9], the solution of the Riemann problem can be obtained by using the fact that the initial jump $\delta \vec{q} = \vec{q}_r - \vec{q}_l$ can be decomposed in terms of the eigenstructure found above in Section 10.10. That is, we wish to find constants α_1 and α_2 such that

$$\alpha_1 \vec{v}_1 + \alpha_2 \vec{v}_2 = \delta \vec{q} . \tag{10.73}$$

This is a matrix problem of the form

$$R \vec{\alpha} = \delta \vec{q} , \tag{10.74}$$

where $R = [\vec{v}_1 \; \vec{v}_2]$. Its solution is

$$\alpha_1 = -\frac{1}{2} \sqrt{\frac{g}{H}} (h_r - h_l) + \frac{1}{2} (u_r - u_l) , \tag{10.75}$$

$$\alpha_2 = \frac{1}{2} \sqrt{\frac{g}{H}} (h_r - h_l) + \frac{1}{2} (u_r - u_l) . \tag{10.76}$$

The terms on the left-hand side of Eq. (10.73) can be interpreted as individual waves $W_1 = \alpha_1 \vec{v}_1$ and $W_2 = \alpha_2 \vec{v}_2$, respectively. The value of $\vec{q}^*$ can be determined by noticing that $\vec{q}^* = \vec{q}_l + W_1$ (or $\vec{q}^* = \vec{q}_r + W_2$). We thus find

$$h^* = \frac{h_l + h_r}{2} + \frac{1}{2} \sqrt{\frac{H}{g}} (u_r - u_l) , \tag{10.77}$$

$$u^* = \frac{u_l + u_r}{2} + \frac{1}{2} \sqrt{\frac{g}{H}} (h_r - h_l) , \tag{10.78}$$

and the full solution is

$$\vec{q}(x,t) = \begin{cases} \vec{q}_l\,, & x < -\sqrt{gH}t\,, \\ \vec{q}^*\,, & -\sqrt{gH}t \le x \le \sqrt{gH}t\,, \\ \vec{q}_r\,, & x > \sqrt{gH}t\,. \end{cases} \tag{10.79}$$

Many numerical methods are concerned with the form of the flux vector in the star region $\vec{F}(\vec{q}^*)$ that is also known as the *numerical flux function*. Here, we find

$$\vec{F}(\vec{q}^*) = \begin{pmatrix} \frac{1}{2}(Hu_l + Hu_r) + \frac{1}{2}\sqrt{gH}\,(h_r - h_l) \\ \frac{1}{2}(gh_l + gh_r) + \frac{1}{2}\sqrt{gH}\,(u_r - u_l) \end{pmatrix}\,, \tag{10.80}$$

which may be re-written as

$$\vec{F}(\vec{q}^*) = \frac{\vec{F}(\vec{q}_l) + \vec{F}(\vec{q}_r)}{2} + \frac{\lambda_2}{2}(\vec{q}_r - \vec{q}_l)\,. \tag{10.81}$$

It turns out that (10.81) can be used in much more general situations and even forms the basis for the Lax-Friedrichs/Rusanov class of approximate Riemann solvers for nonlinear problems for which closed-form solutions become intractable [8,14]. The Lax-Friedrichs/Rusanov flux was also the topic of a Mini-Project #4 in Chapter 9.

10.12 Time-stepping method

The time-stepping technique we apply here to the DG-FEM discretized version of the one-layer model closely follows the "scalar approach" of [4] where splitting is applied such that advective and source terms are time-stepped first, followed by the dispersive terms.

Neglecting the dispersive terms for the time-being since they are not a part of the first splitting step, the method of lines [15] can be applied by noticing that once the DG-FEM integral form (10.6) has been written purely in terms of matrix operators as discussed in the above sections, we recover the system of ordinary differential equations

$$\frac{d\vec{Q}}{dt} = \mathcal{R}(\vec{Q})\,, \tag{10.82}$$

where $\vec{Q} = (h, hu, hv)^\mathsf{T}$ is the vector of unknowns and $\mathcal{R}$ is the DG spatial discretization operator for the advection, Coriolis, and bathymetry source terms. It is assumed that we have left-multiplied the inverse of the local mass matrix, i.e., $\left(\mathcal{M}^k\right)^{-1}$, in arriving at $\mathcal{R}$ so that there is no matrix operator acting on the time-derivative on the left-hand side of (10.82). We have once again followed Eskilsson and Sherwin [4] and time-discretized (10.82) beginning at time-level $t_n = n\Delta t$ using the three-stage

206 CHAPTER 10 Discontinuous Galerkin methods: Background

third-order strong stability preserving Runge-Kutta (SSP-RK) method [12], yielding

$$\vec{V}^{(1)} = \vec{Q}^n + \Delta t \mathcal{R}(\vec{Q}^n) , \tag{10.83}$$

$$\vec{V}^{(2)} = \frac{1}{4} \left(3\vec{Q}^n + \vec{V}^{(1)} + \Delta t \mathcal{R}(\vec{V}^{(1)}) \right) , \tag{10.84}$$

$$\vec{Q}^\dagger = \frac{1}{3} \left(\vec{Q}^n + 2\vec{V}^{(2)} + 2\Delta t \mathcal{R}(\vec{V}^{(2)}) \right) . \tag{10.85}$$

Modal filtering is applied to the spatial discretization operator $\mathcal{R}$ after each stage to help tame aliasing and nonlinearity-driven instabilities. The choice of SSP-RK time-stepper here is not a unique one, and we have mainly used it here since it offers third-order accuracy in time and allows for a simple adaptive time-stepping scheme. That is, Δt can be adjusted after each time-step without changing the coefficients of the scheme. The SSP-RK methods have gained favor in the DG-FEM literature [2,8] since it guarantees no oscillations are introduced as a result of time-stepping for problems involving discontinuities and shocks. As mentioned previously, however, shocks and discontinuous features are not of concern for the equations under consideration here since the dispersive term will "slow down" any short waves that would contribute to the formation of a shock, leading instead to more physically realistic features such as nonlinear wavetrains or solitary waves.

As discussed in Section 10.9, the next step in the "scalar approach" is to solve the wave continuity equation. Its continuous form is given by

$$\nabla \cdot \left(\gamma \nabla z^\dagger \right) - z^\dagger = -\nabla \cdot \vec{a}^\dagger , \tag{10.86}$$

with $\gamma = H^2/6$ and $\vec{a}$ as defined previously by Eq. (10.55). The spatially discretized vector $\vec{a}^\dagger$ can be computed quite simply by evaluating $\mathcal{R}(\vec{Q}^\dagger)$ for only the **hu** and **hv** equations. The auxiliary variable $z^\dagger$ is then computed by inverting the matrix representation of the SIP-DG formulation of the Helmholtz operator (10.63). The momentum equations are then updated via

$$(h\vec{u})^{n+1} = (h\vec{u})^\dagger + \gamma \Delta t \nabla z^{n+1} , \tag{10.87}$$

where the DG-FEM discretization of the source terms involving nonlinear products with gradients of known quantities is discussed in Section 10.8. Hence, the vector of unknowns at time t_{n+1} is updated via $\vec{Q}^{n+1} = (h^\dagger, hu^{n+1}, hv^{n+1})^\mathsf{T}$.

10.13 Mini-projects

1. Show that in 1D the following expression for the upwind-flux reduces to cu_L for $c > 0$ flow and cu_R for $c < 0$ flow:

$$(cu)^* = \{\!\{cu\}\!\} + \frac{1}{2}|c|[\![u]\!] . \tag{10.88}$$

Here. u_L (u_R) corresponds to the value of the function u on the left (right) of an interface. *Hint: Draw a sketch of the line segment corresponding to a particular interior element with its two edges. Starting with $c > 0$ flow, what are the two cases to consider?*

2. Consider a general element $\vec{D}^k = [x_k, x_{k+1}]$ for some particular $k \in \{0, 1, \cdots, K\}$. Write down a linear transformation $x = T(r)$ that maps the standard bi-unit interval $[-1, 1]$ to element $\vec{D}^k$. Then solve for the inverse transformation $r = T^{-1}(x)$. For a general function $f(x)$ that is defined everywhere, express its first derivative $f'(x)$ in $\vec{D}^k$ in terms of the $r \in [-1, 1]$ variable. *Hint: Use the chain rule.*

3. Provide a derivation of the DG formulation for the 1D tracer (linear advection) equation

$$\frac{\partial T^k}{\partial x} + c\frac{\partial T^k}{\partial x} = 0, \tag{10.89}$$

by multiplying by a Lagrange interpolant (test function) $\ell_j(x)$, integrating over a particular element (sub-interval) $\mathbf{D}^k$, and substituting in the nodal expansion for u. Simplify your equality by introducing local mass and stiffness matrices where appropriate. You need not substitute a numerical flux function for the "starred" variable.

4. Consider the quadratic polynomials $f(x) = x^2 - 3x + 7$ and $g(x) = 21x^2 + 6x - 5$. Calculate the quartic polynomial $h(x) = f(x)g(x)$. Using a software tool of your choice, compute the point-wise product fg at 16 equally-spaced grid points covering the interval $[-1, 1]$. Plot the results with a computer plotting tool, or on graph paper. Are the two products the same, slightly different, or significantly different? Explain why using concepts from this chapter.

5. Consider a long rectangular transparent water tank with a vertical dividing wall placed in the middle of the long sides of the tank, splitting it into left and right halves. The left side of the tank is filled with water up to a height of 10 cm and the right side of the tank is filled with water up to a height of 14 cm. Sketch this setup in 1D. Suppose at $t = 0$ the dividing wall is instantaneously removed. Using the linear Riemann problem discussed in this chapter, sketch the free surface at $t = 0.2$ s. Calculate the u field at the same time and sketch it. Is this what you would see in a lab experiment? If yes, explain why. If no, explain how you could improve your analysis (you don't need to carry-out another analysis).

References

[1] A.S. Almgren, J.B. Bell, W.Y. Crutchfield, Approximate projection methods: Part I. Inviscid analysis, SIAM Journal on Scientific Computing 22 (4) (2000) 1139–1159.

[2] B. Cockburn, C.W. Shu, TVB Runge-Kutta local projection discontinuous Galerkin finite-element method for conservation-laws. 2. General framework, Mathematics of Computation 52 (186) (1989) 411–435, https://doi.org/10.2307/2008474.

208 **CHAPTER 10** Discontinuous Galerkin methods: Background

[3] Colin J. Cotter, et al., LBB stability of a mixed Galerkin finite element pair for fluid flow simulations, Journal of Computational Physics 228 (2) (2009) 336–348, https://doi.org/10.1016/j.jcp.2008.09.014.

[4] C. Eskilsson, S. Sherwin, Spectral/hp discontinuous Galerkin methods for modelling 2D Boussinesq equations, Journal of Scientific Computing 22 (2005) 269–288.

[5] E. Ferrer, R.H.J. Willden, A high order discontinuous Galerkin finite element solver for the incompressible Navier-Stokes equations, Computers & Fluids 46 (2011) 224–230.

[6] F.X. Giraldo, The Lagrange-Galerkin spectral element method on unstructured quadrilateral grids, Journal of Computational Physics 147 (1998) 114–146.

[7] W.N. Gordon, C.A. Hall, Construction of curvilinear coordinate systems and application to mesh generation, International Journal for Numerical Methods in Engineering 7 (1973) 461–477.

[8] J. Hesthaven, T. Warburton, Nodal Discontinuous Galerkin Methods, Springer, 2008.

[9] R.J. Leveque, Finite Volume Methods for Hyperbolic Problems, Cambridge University Press, 2002.

[10] R. Peyret, Spectral Methods for Incompressible Viscous Flow, Springer-Verlag New York, Inc., 2002.

[11] K. Shahbazi, P.F. Fischer, C.R. Ethier, A high-order discontinuous Galerkin method for the unsteady incompressible Navier-Stokes equations, Journal of Computational Physics 222 (2007) 391–407.

[12] C.W. Shu, S. Osher, Efficient implementation of essentially non-oscillatory shock-capturing schemes, Journal of Computational Physics 77 (1988) 439–471.

[13] D.T. Steinmoeller, M. Stastna, K.G. Lamb, A short note on the discontinuous Galerkin discretization of the pressure projection operator in incompressible flow, Journal of Computational Physics 251C (2013) 480–486.

[14] E.F. Toro, Riemann Solvers and Numerical Methods for Fluid Dynamics, 2nd ed., Springer, 1999.

[15] L.N. Trefethen, Spectral Methods in MATLAB, Society for Industrial and Applied Mathematics, 2000.

[16] T. Warburton, An explicit construction of interpolation nodes on the simplex, Journal of Engineering Mathematics 56 (2006) 247–262.

CHAPTER

Modeling fluid transport processes with discontinuous Galerkin methods: Implementation

11

CONTENTS

11.1 Towards practical implementations of DG-FEM 209
11.2 Spurious eddies in inviscid DG-FEM solutions 210
11.3 Curvilinear elements 214
11.4 Constructing coordinates systems for curvilinear elements 214
11.5 Cubature and quadrature integration 217
11.6 Internal rotating seiche simulation using curvilinear elements 219
11.7 Internal rotating seiche simulation in a real-world lake 220
11.8 Wrap-up 223
References 223

11.1 Towards practical implementations of DG-FEM

In Chapter 10, we have presented the relevant aspects of the theory of Discontinuous Galerkin methods for the dispersion modified shallow water equations. The intention was to provide a complete picture for the small subset of readers interested in, what is in all honesty, a lengthy, and rather unpleasant exercise of mathematical notation. However, as with much of modern applied mathematics, this notation is necessary to make any headway in reading the active literature in this field. This chapter, takes a different point of view, by focusing on how the implementation of the theory to an actual basin almost immediately requires a modification. In effect, we are doing something the scientific literature is not good at: we are showcasing a failure, and only then do we derive the means to improve the method. To be more concrete, the necessity of curvilinear elements for general situations is illustrated due to singular/spurious flow features that can emerge due to a piece-wise linear representation of the boundary when used alongside an element-wise modal filter. In a nut-shell, the damping effect of the modal filter plays the role of viscosity in real-world flows, by diffusing the pressure-drop singularity away from the boundary, allowing for the flow to 'detach' from flowing along the headland. However, the local modal filter was never intended to provide such a process, and the numerical 'result' is spurious/unwanted! Following the demonstration of these numerical artifacts, it is then explained

Physics and Ecology in Fluids. https://doi.org/10.1016/B978-0-32-391244-0.00021-8
Copyright © 2023 Elsevier Inc. All rights reserved.

209

CHAPTER 11 Discontinuous Galerkin methods: Implementation

how the nodal DG-FEM method should be extended using high-order cubature and quadrature integration rules to deal with the non-constant geometric (i.e., Jacobian) factors introduced by curvilinear elements providing further numerical stabilization and suppression of aliasing artifacts. Highlight simulations of a mid-sized lake are provided alongside these developments.

For the more practically-minded reader looking to 'cut to the chase' and perhaps less interested in the model development and improvement process, we list here the three python codes that may be used to reproduce the simulations presented herein:

1. `nhsw2d_dg_peninsula_straight.py`
2. `nhsw2d_dg_peninsula_curved.py`
3. `nhsw2d_dg_Pinehurst_Lake_curved.py`

11.2 Spurious eddies in inviscid DG-FEM solutions

As one begins to explore outside of traditional geometries with the DG-FEM solver developed in the previous chapter, it would be discovered that under certain conditions spurious eddies, corresponding to an unphysical production of vorticity, appear to form in the domain near obstacles resembling re-entrant corners (i.e., boundary corners that protrude inwards into the domain). This effect is illustrated in Fig. 11.1 where our annular basin has been perturbed to include a peninsula. The DG-FEM solver with polynomial order $N = 4$ was initialized with the initial conditions

$$\eta = 3.2(x/8000),\qquad(11.1)$$
$$u = 0,\qquad(11.2)$$
$$v = 0,\qquad(11.3)$$

corresponding to an at-rest tilt in the free surface field.

A numerical instability occurred shortly after $t = 27$ h, preventing further time-stepping, though the reason for the instability was evident earlier due to the sharp gradients visible near the corner in Fig. 11.1.

The eddies bear a striking resemblance to boundary-layer separation eddies that would occur due to flow past an obstacle in viscous flow [5]. However, since our model equations do not contain any viscous terms, the formation of a viscous boundary-layer is not possible and hence boundary-layer separation should not be possible. These spurious eddies are thus discretization artifacts, and appear to coincide with the presence of a sharp re-entrant corner. Even in the cases where the actual boundary is smooth, re-entrant corners at the element-scale may result as a consequence of the piece-wise linear representation of the boundary assumed in mesh generation. Although these artifacts are spurious in the sense that inviscid flow around an obstacle should not separate, from a theoretical stand-point they should be expected. Below, we explain why this is the case and propose methods for remedying the situation.

11.2 Spurious eddies in inviscid DG-FEM solutions

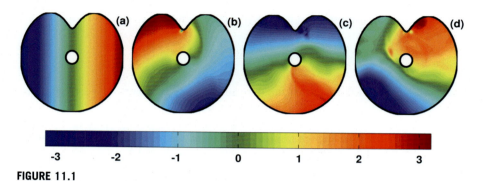

FIGURE 11.1

Snapshots of the η-field in the order $N = 4$ DG-FEM simulation of a rotating seiche on a perturbed circular domain with a re-entrant peninsula at **(a)** $t = 0$ h, **(b)** $t = 6.8$ h, **(c)** $t = 14.0$ h, **(d)** $t = 20.9$ h. Note the apparent separation eddies visible near the peninsula in panels **(b)**–**(d)**. This figure was published in *Ocean Modelling*, **107**, D.T. Steinmoeller, M. Stastna, K.G. Lamb, "Discontinuous Galerkin methods for dispersive shallow water models in closed basins: Spurious eddies and their removal using curved boundary methods," 112–124. © 2016 Elsevier.

Consider the simpler situation of inviscid, incompressible, and irrotational (i.e., potential) flow. That is, we assume

$$\nabla \cdot \vec{u} = 0, \quad \nabla \times \vec{u} = 0. \tag{11.4}$$

In general, these constraints will certainly not be true during simulations and we presently make this assumption simply to gain insight from theory. Under these constraints, the velocity field has both a velocity potential and a streamfunction, i.e.,

$$\vec{u} = \nabla \phi = \nabla^\perp \psi, \tag{11.5}$$

where $\nabla^\perp = \hat{k} \times \nabla = \left(-\frac{\partial}{\partial y}, \frac{\partial}{\partial x}\right)$. Suitably differentiating and adding the various components (or by using vector calculus identities), it is straightforward to show that ϕ and ψ both satisfy Laplace's equation

$$\nabla^2 \phi = 0, \quad \nabla^2 \psi = 0. \tag{11.6}$$

The no-normal flow boundary conditions imply that

$$\frac{\partial \phi}{\partial n} = 0, \quad \text{on} \quad \partial \Omega, \tag{11.7}$$

$$\psi = \text{constant}, \quad \text{on} \quad \partial \Omega. \tag{11.8}$$

It is useful to define a complex velocity potential

$$w(z) = \phi + i\psi, \tag{11.9}$$

212 **CHAPTER 11** Discontinuous Galerkin methods: Implementation

where $z = x + iy$, from which the velocity components (u, v) can be recovered by noticing that

$$\frac{dw}{dz} = \frac{\partial(\phi + i\psi)}{\partial x} = u - iv,\tag{11.10}$$

where we have used the fact that the derivative of any complex-valued function is independent of the direction taken in the xy-plane.

Progress was made in the area of potential flow theory not by solving Laplace's equation, but rather by using the result from complex analysis that any analytic function (a complex-valued function whose derivative exists) has real and imaginary parts that satisfy Laplace's equation [5]. Thus, any analytic function that is known represents a potential flow solution for some particular situation. What type of obstacle the fluid is flowing around depends on properties of the function, $w(z)$, itself.

It is known that the complex potential for flow around a wall angle $\alpha = \pi/n$ is given by [5]:

$$w = Az^n, \quad \left(n \geq \frac{1}{2}\right).\tag{11.11}$$

In polar coordinates, $z = re^{i\theta}$,

$$w = A\left(re^{i\theta}\right)^n = Ar^n(\cos(n\theta) + i\sin(n\theta)),\tag{11.12}$$

hence

$$\phi = Ar^n\cos(n\theta), \quad \psi = Ar^n\sin(n\theta).\tag{11.13}$$

Noticing that $\psi = 0$ for $\theta = 0, \pi/n$ and using the fact that any streamline along which $\psi = constant$ can represent a boundary, we recognize that this particular complex potential corresponds to flow around a sharp corner with angle $\theta = \pi/n$.

At this point, we can recover the velocity using Eq. (11.10)

$$u = \Re\left(nAz^{n-1}\right) = \Re\left(\frac{A\pi}{\alpha}z^{\frac{\pi-\alpha}{\alpha}}\right),\tag{11.14}$$

$$v = \Im\left(nAz^{n-1}\right) = \Im\left(\frac{A\pi}{\alpha}z^{\frac{\pi-\alpha}{\alpha}}\right).\tag{11.15}$$

Paying close attention to the expression on the right-hand side, it is clear that the exponent to which z is raised is negative whenever the wall-angle exceeds π, and thus the velocity is not defined when $z = 0$ (the corner).

Even in the case where $0 < \alpha < \pi$, the point $z = 0$ is a singular point since the velocity's derivatives do not exist. We thus reach the main point of this discussion, that is also highlighted by Kundu [5] and is depicted in Fig. 11.2: the velocity at a wall-corner is infinite if the wall-angle is greater than $180°$ and is zero for wall-angles less than $180°$. Thus, near re-entrant corners the numerical solution should be *expected* to be poorly behaved since the exact potential flow solution is also. Although the velocity derivatives do not exist at corners less than 180°, this does not appear to be an issue for the numerical solution.

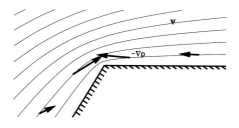

FIGURE 11.2

Cartoon diagram of potential flow around a wall corner of angle $\alpha > \pi$. Contours depict streamlines (lines where $\psi = const.$) and arrows illustrate the relative strength and direction of $-\nabla p$.

The "spurious eddies" encountered in simulations begin as very steep free-surface depressions that diffuse away from the boundary, to understand this phenomenon, consider Bernoulli's equation

$$p + \frac{1}{2}\rho_0 \|\mathbf{u}\|^2 = const., \qquad (11.16)$$

where p is the pressure. Taking the gradient and re-arranging gives

$$-\nabla p = \frac{1}{2}\rho_0 \nabla \left(\|\mathbf{u}\|^2\right). \qquad (11.17)$$

Since the flow is infinite at the corner corresponding to $z = 0$, the upstream fluid must accelerate as it approaches the corner and decelerate as it moves away from it. This fact, together with Eq. (11.17), implies that $-\nabla p$ points in the direction with the flow upstream of the corner (i.e., a favorable pressure gradient) and *against* the flow downstream of the corner (i.e., an adverse pressure gradient). Hence, a region of very low pressure must exist in the vicinity of the corner. It is thus clear then why the singularity would manifest itself as a large (localized) free-surface depression since in the shallow water framework $p = \rho_0 g \eta$. In general, we found that the free-surface depression is advected downstream of the corner immediately after appearing, since the numerical method cannot compensate for the singularity.

In real-world flow around a corner, the region of adverse pressure gradient would cause the flow to separate from the corner resulting in the formation of eddies due to effects in the viscous boundary layer [5]. In the DG-FEM simulations discussed, the observed eddies are a result of the local modal filtering that attempts to stabilize the pressure singularity by diffusing it away from the boundary, taking over the role of viscosity in realistic flows. This effect of the filter was discovered by turning off the filter and observing singular growth at the corner that leads to numerical blow-up with no eddy introduced. It was also found that spurious eddy generation is more

214 **CHAPTER 11** Discontinuous Galerkin methods: Implementation

prominent in simulations where nonlinear effects are non-negligible. The fact that a standard filter coupled with the presence of re-entrant corners will typically lead to spurious eddies is a dangerous feature of the numerical model, since a modeler may be led to believe that these eddies are physical, when in fact they are the result of the filter's action on a part of the solution that is singular. For instance, in [7], spurious eddies due to a limiter are presented as physical for the situation of supersonic compressible flow past an equilateral triangle. Despite the effort of filtering, it has been found that this singular behavior can still lead to numerical blow-up. Thus, some effort must be taken to remedy this problem, as discussed in the next section.

11.3 Curvilinear elements

In addition to solution singularities, it is also known that the convergence rates of a high-order method may be limited to sub-optimal rates as a result of an inaccurate representation of the boundary. This fact was demonstrated in [4] who demonstrated poor convergence rates for the solution to Maxwell's equations on a circular domain with a piece-wise linear representation of the boundary. Dupont [1] has also suggested that rounding singular corners is necessary to suppress poor polynomial behavior resulting from the high-order DG-FEM in his inter-model comparison of the oceanic shallow water equations. It is thus apparent that a high-order method begs for a smooth and accurate representation of the boundary, and hence, deformed or curvilinear elements along the boundary will be necessary to achieve accurate solutions on general lake geometries with the high-order discontinuous Galerkin method.

11.4 Constructing coordinates systems for curvilinear elements

We have adopted the approach in [4] that avoids some of the difficulties and cumbersome work associated with explicitly constructing two-dimensional mapping functions for high-order curvilinear elements, e.g., explicitly calculating high-order "shape-functions". The technique discussed here generalizes well to elements with an arbitrary number of nodes and thus allows for the robust construction of high-order curvilinear elements. The method discussed here represents an extension of the technique used in [4] for circular boundaries, since we consider arbitrary domain boundaries represented by cubic splines of B-spline type.

Assume we have generated a straight-sided finite element mesh that approximates the boundary in a piece-wise linear manner, and also assume we have retained a smooth representation of the boundary in a parameterized curve $\mathcal{C}: \vec{x}_b(t) = (x_b(t), y_b(t))$ (see Fig. 11.3), that we assume to be parameterized by arc-length $0 \le t \le S$. In practice, we have found taking $\mathcal{C}$ to be a parametric cubic-spline in-

11.4 Constructing coordinates systems for curvilinear elements

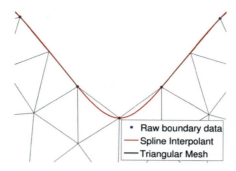

FIGURE 11.3

Illustration of straight-sided element mesh along with a smooth representation of the boundary, the cubic spline interpolant, that will be used to produce deformed elements. This figure was published in *Ocean Modelling*, **107**, D.T. Steinmoeller, M. Stastna, K.G. Lamb, "Discontinuous Galerkin methods for dispersive shallow water models in closed basins: Spurious eddies and their removal using curved boundary methods," 112–124. © 2016 Elsevier.

terpolant (of B-spline type) of the boundary to be a simple and effective choice. The algorithm for a particular element that is to be curved is as follows:

1. Adjust the straight-sided finite element mesh by moving the vertices (i.e., end-points only) of the straight-sided element's boundary edge e so that they lie exactly at points on $\mathcal{C}$, say $\vec{x}_b(t_1)$ and $\vec{x}_b(t_2)$.
2. Distribute the 1D LGL nodes along the curved edge by arc-length using the parameterization $\vec{x}_b(t)$ for $t_1 \leq t \leq t_2$ to obtain new local coordinates along the curved edge, denoted $\vec{x}_{curved}(r,s)|_e$, where (r,s) are the coordinates of the reference triangle (see Fig. 11.4).
3. Calculate the deformation (displacement field) in moving from the edge nodes from the straight edge to the curve $\mathcal{C}$, i.e., $\vec{w}(r,s) = \vec{x}_{curved}(r,s)|_e - \vec{x}_{straight}(r,s)|_e$, also called the *warp* factor.
4. "Blend" the edge deformation to the interior nodes using Gordon-Hall blending (see below) to obtain new local coordinates for the whole element: $\vec{x}_{curved}(r,s) = \vec{x}_{straight}(r,s) + b(r,s)\vec{w}(r,s)$, where $b(r,s)$ is a *blending* function.
5. Compute local metric factors, i.e., x_r, y_r, x_s, y_s, and Jacobian $J = x_r y_s - x_s y_r$, numerically using the differentiation matrices on the reference element $\mathcal{D}_r$, $\mathcal{D}_s$.

The one point that requires further attention is how to choose a blending function $b(r,s)$ to appropriately "blend" the edge deformation on the element boundary to the interior of the element. To motivate our discussion, consider the simplistic one-dimensional case where two function values f_0 and f_1 are known at points x_0 and x_1 and we wish find a function $f(x)$ to interpolate to points inside the interval $[x_0, x_1]$. If, for additional simplicity, we assume $f_0 = 0$, we realize that the only way to interpolate to interior points with the information that we have is by the linear Lagrange interpolant $\ell_1(x) = (x - x_0)/(x_1 - x_0)$, i.e.,

$$f(x) = \left(\frac{x - x_0}{x_1 - x_0}\right) f_1. \qquad (11.18)$$

In a sense, we have found the appropriate blending function to be $\ell_1(x)$ since this function satisfies the desired properties: $\ell_1(x_1) = 1$, $\ell_1(x_0) = 0$.

Now consider the two-dimensional case where, for example, our edge deformation $\vec{w}(r, s)$ is known along the triangle edge corresponding to the line $r = -1$ for $-1 \leq s \leq 1$ on the reference element (Fig. 11.4). Clearly, we require the blending function to satisfy $b(r = -1, s) = 1$ since this is the only region where information is known. It also seems sensible that the effect of the edge-deformation would decay to zero at the opposite triangle edge that lies on the line $s = -r$ for $-1 \leq r \leq 1$, leading us to define the blending function as

$$b(r, s) = \left(\frac{s + r}{s - 1}\right), \qquad (11.19)$$

that satisfies $b(r = -1, s) = 1$ and $b(r, s = -r) = 0$, as required. The one issue that remains is the apparent singularity at the point $(-1, 1)$. This point corresponds to a location where $\vec{w} = 0$ since it is a vertex of the finite element mesh that does not need to be deformed. Thus, we can simply apply the blending at nodal points not corresponding to the singular point in step 4 above.

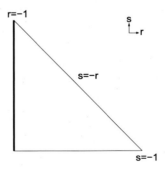

FIGURE 11.4

Diagram of the reference triangle and illustration of (r, s) coordinates. This figure was published in *Ocean Modelling*, **107**, D.T. Steinmoeller, M. Stastna, K.G. Lamb, "Discontinuous Galerkin methods for dispersive shallow water models in closed basins: Spurious eddies and their removal using curved boundary methods," 112–124. © 2016 Elsevier.

While the "blending" procedure discussed is a straight-forward extension of linear Lagrange interpolation to two-dimensions, one subtle difference between Lagrange interpolation is that the two-dimensional blending function is chosen to be zero or one along entire *line segments*, and not at points in space. It is for this reason that the technique has been referred to as "transfinite interpolation" [3], since in general

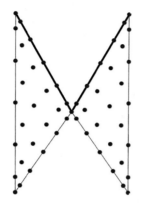 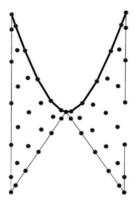

FIGURE 11.5

Left: A pair of elements before being deformed. **Right**: The same elements after being deformed to match the cubic-spline representation of the boundary with interior nodes redistributed via Gordon-Hall blending. This figure was published in *Ocean Modelling*, **107**, D.T. Steinmoeller, M. Stastna, K.G. Lamb, "Discontinuous Galerkin methods for dispersive shallow water models in closed basins: Spurious eddies and their removal using curved boundary methods," 112–124. © 2016 Elsevier.

the data is being sampled over a continuum and not just at a finite set of points. An example of a pair of elements that have undergone the full "deform and blend" transformation are shown in Fig. 11.5.

11.5 Cubature and quadrature integration

The nodal approach described in the above sections relies heavily on the assumption that the Jacobian of the mapping from a particular element to the standard element is a constant, and hence may be brought outside of the integrals in the nodal DG-FEM formulation. Once we have introduced curvilinear elements, the Jacobian on these elements is no longer constant and we must thus pay a computational price. Firstly, a separate mass matrix must be stored for each curvilinear element, thereby driving up computational storage costs. Secondly, the Jacobians of the mappings used here are rational functions of the standard element's coordinates, and their product with solution fields will in general lead to aliasing errors that can drive numerical instabilities.

Nonlinearities involving rational functions cannot be de-aliased completely (as, for example, a quadratic nonlinearity could) since their polynomial representation would consist of a Taylor series with infinitely many terms. Nevertheless, a great deal of aliasing error can be removed by evaluating the integrals with cubature rules that are of higher order than the approximating polynomials. Here, "cubature" refers to the higher-dimensional analogy to 1D quadrature rules. For polynomials of order

218   CHAPTER 11   Discontinuous Galerkin methods: Implementation

N, we follow [4] and evaluate these integrals with cubature rules for the reference triangle of order $3(N + 1)$. A general inner-product of two functions f and g are evaluated using a cubature rule by

$$\int_{\mathbf{D}^k} fg\, d\mathbf{x} \approx \sum_{i=1}^{N_c} f(\mathbf{r}_i^c) g(\mathbf{r}_i^c) J_i^k w_i^c , \tag{11.20}$$

where J_i^k is the Jacobian of the mapping from the standard element D^k, w_i^c are the cubature weights associated with cubature nodes $\{\mathbf{r_i^c}\}_{i=1}^{N_c}$. The cubature nodes and weights are provided by the symmetric rules in [6] and implemented in MATLAB® in [4].

The use of cubature integration makes the evaluation of the local mass and stiffness matrices more computationally expensive, since additional interpolation operations must be carried out to interpolate integrands to the cubature nodes. In particular, we define the $N_c \times N_p$ interpolation matrix $V_{ij}^c = \ell_j(\mathbf{r}_i^c)$ to interpolate functions defined at the polynomial interpolation nodes to the cubature nodes. The $N_p \times N_p$ mass matrix can then be found using cubature integration as follows

$$\mathcal{M}_{lm}^k = \int_{\mathbf{D}^k} \ell_l^k(\vec{x}) \ell_m^k(\vec{x})\, d\mathbf{x} \tag{11.21}$$

$$\approx \sum_{i=1}^{N_c} \ell_l(\mathbf{r}_i^c) \ell_m(\mathbf{r}_i^c) w_i^c J_i^{k,c} . \tag{11.22}$$

Hence,

$$\mathcal{M}^k = \left(\mathcal{V}^c\right)^{\mathsf{T}} \mathcal{W}^k \mathcal{V}^c , \tag{11.23}$$

where $\mathcal{W}^k$ is the $N_c \times N_c$ diagonal matrix with entries $\mathcal{W}_{ii}^k = w_i^c J_i^{k,c}$. For the local stiffness matrix,

$$\mathcal{S}_{x,nm}^k = \int_{\mathbf{D}^k} \ell_n^k(\vec{x}) \frac{\partial \ell_m^k}{\partial x}(\vec{x})\, d\mathbf{x} \tag{11.24}$$

we must invoke the chain rule to express the operators in terms of the $N_p \times N_p$ differentiation matrices on the reference triangle, $\mathcal{D}_r$ and $\mathcal{D}_s$, yielding

$$\mathcal{S}_x^k = \left(\mathcal{V}^c\right)^{\mathsf{T}} \mathcal{W}^k \left(\mathrm{diag}(r_x^k(\mathbf{r}_i^c)) \mathcal{V}^c \mathcal{D}_r + \mathrm{diag}(s_x^k(\mathbf{r}_i^c)) \mathcal{V}^c \mathcal{D}_s\right) . \tag{11.25}$$

An identical argument gives

$$\mathcal{S}_y^k = \left(\mathcal{V}^c\right)^{\mathsf{T}} \mathcal{W}^k \left(\mathrm{diag}(r_y^k(\mathbf{r}_i^c)) \mathcal{V}^c \mathcal{D}_r + \mathrm{diag}(s_y^k(\mathbf{r}_i^c)) \mathcal{V}^c \mathcal{D}_s\right) . \tag{11.26}$$

In addition to volume (two-dimensional) integrals, surface integral (element-coupling) terms must also be computed using Gaussian quadrature, with analogous two-dimensional interpolation operators used to evaluate the integrand at the appropriate quadrature points along an edge. We again follow [4] and use order $N_G = 2(N+1)$ Gaussian quadrature along the edges.

11.6 Internal rotating seiche simulation using curvilinear elements

We now consider the same simulation shown in Section 11.2 where a circular basin has been perturbed to include a peninsula. The difference here is that we employ the developments on curvilinear elements described in the above sections along with polynomial order $N = 8$. All boundary elements have been deformed such that their boundary edges conform to a cubic spline interpolant (of B-spline type) of the boundary.

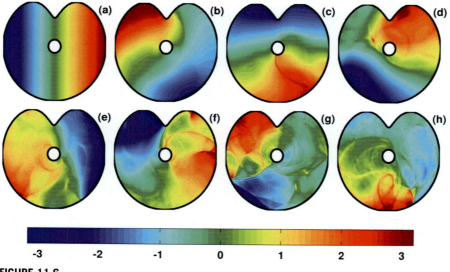

FIGURE 11.6

Panels **(a)**–**(d)**: Like Fig. 11.1 but with curvilinear elements along the boundary. The other panels correspond to the later times **(e)** $t = 28.1$ h, **(f)** $t = 34.9$ h, **(g)** $t = 42.1$ h, **(h)** $t = 49.0$ h. This figure was published in *Ocean Modelling*, **107**, D.T. Steinmoeller, M. Stastna, K.G. Lamb, "Discontinuous Galerkin methods for dispersive shallow water models in closed basins: Spurious eddies and their removal using curved boundary methods," 112–124. © 2016 Elsevier.

The results of the simulation are shown in Fig. 11.6. In addition to finding that the simulation is apparently long-term stable, unlike in the straight-sided case, we also note that the spurious eddies associated with the sharp re-entrant corner have been suppressed since the peninsula is now represented in a geometrically smooth way.

It is important to note that although the spurious eddies have been suppressed, the region of the flow at the tip of the peninsula still represents a geometric feature where a strong adverse pressure gradient must appear in order to decelerate flow around the obstacle. Indeed, an adverse η-gradient appears between $t = 2.5$ h and $t = 5.4$ h (not shown). It is possible that such high gradients in the physical fields can still lead to instabilities, as one can determine through numerical experimentation.

As discussed in Section 11.2, in real-world flows it is certainly reasonable to expect the flow around the peninsula to separate and generate eddies due to viscous boundary-layer effects, but since we have not included a physical model for such processes we are left in the somewhat precarious situation in which we demand the flow to remain "attached" to the peninsula in all cases.

11.7 Internal rotating seiche simulation in a real-world lake

In this section, we provide proof-of-concept that the high-order DG-FEM methodology of this chapter can be applied to real-world lake geometries involving irregular coastlines. Bathymetry data at a resolution of 50 m for the mid-sized Pinehurst Lake, Alberta has been obtained from the Alberta Geological Survey website:

```
http://www.ags.gov.ab.ca/
```

Pinehurst Lake was chosen since its data set was freely available and its size is such that both rotation and stratification effects are expected to be important in the summer months. The raw data consists of a Cartesian grid with 216×245 data points containing both land and water measurements. A plot of the 50 m bathymetry data is shown in Fig. 11.7 where land values have been set to zero.

A parametric representation of the coastline was obtained using the data returned by MATLAB's `contour` function used to obtain the zero-depth contour and is shown in Fig. 11.7(**b**). It was found that finite element meshes generated from the raw data contained $O(10,000)$ elements and possessed poor mesh quality (i.e., large aspect ratio triangles and large element size gradients) since the raw 0-depth contour is far from smooth. A smoothed piece of coastline is shown in Fig. 11.7(**c**) with corresponding $N = 6$ curved finite element mesh in panel (**d**). The smoothed coastline was found by convolving the two-dimensional bathymetry data with the 2D cardinal B-spline 16 times and sub-sampling the result to a 200 m resolution data set. A piece-wise cubic spline interpolant (of B-spline type) of the coastline was then constructed so that boundary elements could be deformed using the techniques explained in Section 11.3. The straight-sided finite element mesh, that is later deformed by our DG-FEM solver, was constructed using the open-source `gmsh` software [2]. Finally, the depth-profile $H(x, y)$ was linearly interpolated from the Cartesian data to our unstructured DG-FEM mesh for use during simulations. The depth-profile $H(x, y)$ was capped at a minimum depth of 6 m to avoid dry states that would drive the DG-FEM solver unstable.

11.7 Internal rotating seiche simulation in a real-world lake

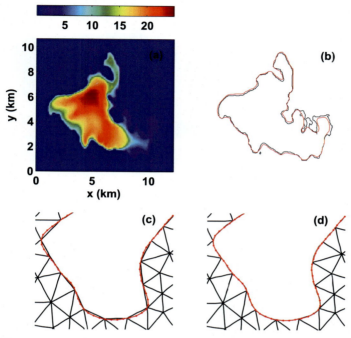

FIGURE 11.7

Panel **(a)**: Depth (in m) of Pinehurst Lake, AB from raw 50 m bathymetry data, and panel **(b)**: corresponding $H = 0$ contour (black) with smoothed coastline super-imposed (red). The lower panels show a zoomed-in section of the **(c)** straight-sided and **(d)** curved ($N = 6$) finite element mesh with $K = 1807$ elements near $(x, y) = (7 \text{ km}, 5 \text{ km})$ with cubic spline interpolant super-imposed (red). This figure was published in *Ocean Modelling*, **107**, D.T. Steinmoeller, M. Stastna, K.G. Lamb, "Discontinuous Galerkin methods for dispersive shallow water models in closed basins: Spurious eddies and their removal using curved boundary methods," 112–124. © 2016 Elsevier.

As in previous simulations, the reduced gravity is $g' = (\Delta\rho/\rho_0)g = 0.024525$ m s^{-2}, where $(\Delta\rho/\rho_0) = 0.0025$. The Coriolis parameter was taken to be $f = 1.1863 \times 10^{-4}$ s^{-1}, corresponding to the 54.65° latitude of Pinehurst Lake. Results of an $N = 6$ DG-FEM simulation from an initial east-west interfacial tilt taken to increase linearly from $\eta = 0$ to $\eta = 2.5$ m are shown in Fig. 11.8 that illustrates the evolving density interface at fixed-time snapshots with the initial condition plotted in panel **(a)**. Since the relative amplitude of the initial condition compared to the depth is, on average, not as large as in previous simulations in this chapter, nonlinear effects are expected to be weaker. In spite of this fact, panels **(c)** and **(d)** show that nonlinear (high-amplitude) waves emerge in the shallows in the southeastern part of the basin after sufficient time has passed. As a result, small scale waves have proliferated throughout the entire basin by $t = 62.7$ h.

222 CHAPTER 11 Discontinuous Galerkin methods: Implementation

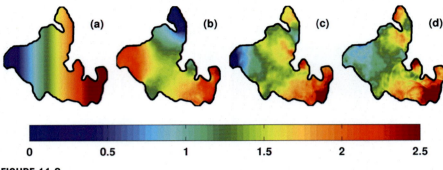

FIGURE 11.8

Evolution of an interfacial tilt in Pinehurst Lake, AB using the $N = 6$ DG-FEM with curvilinear boundary elements at times **(a)** $t = 0$ h, **(b)** $t = 19.4$ h, **(c)** $t = 39.3$ h, **(d)** $t = 62.7$ h. This figure was published in *Ocean Modelling*, **107**, D.T. Steinmoeller, M. Stastna, K.G. Lamb, "Discontinuous Galerkin methods for dispersive shallow water models in closed basins: Spurious eddies and their removal using curved boundary methods," 112–124. © 2016 Elsevier.

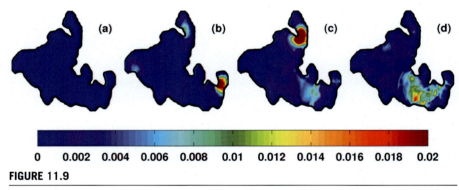

FIGURE 11.9

Like Fig. 11.8, except the kinetic energy density, $\frac{1}{2}h(u^2 + v^2)$ is plotted. This figure was published in *Ocean Modelling*, **107**, D.T. Steinmoeller, M. Stastna, K.G. Lamb, "Discontinuous Galerkin methods for dispersive shallow water models in closed basins: Spurious eddies and their removal using curved boundary methods," 112–124. © 2016 Elsevier.

Fig. 11.8 should be compared closely to Fig. 11.9 where the kinetic energy density is plotted at the same times. At the earlier times (panels **(b)** and **(c)**), the most energetic features correspond to attached flow around peninsulas or other coastal obstacles. It is apparent that geometric focusing intensifies such features when they occur in narrow, confined parts of the basin. Panel **(d)** illustrates the kinetic energy fingerprint of small scale internal wave activity localized in the shallow eastern end of the lake at later times.

11.8 Wrap-up

In this chapter, we have learned that DG-FEM at mid- to high-orders with a piecewise linear representation of the domain boundary is far from flawless. Indeed, the so-called "high-order," i.e., higher than linear, approximating polynomials in essence *demand* that the boundary and background data be *de facto* sufficiently smooth features to avoid the appearance of unwanted oscillations and artifacts. Indeed, a common theme in applied mathematics in general is the notion that non-smooth data and smooth approximating functions do not mix. See, for instance, the infamous *Gibbs phenomenon* of Fourier (periodic signal component) analysis or the associated *Runge phenomenon*, Gibbs' cousin for orthogonal polynomials:

```
https://en.wikipedia.org/wiki/Gibbs_phenomenon
https://en.wikipedia.org/wiki/Runge%27s_phenomenon
```

To overcome the obstacles summarized in the above paragraph, concepts such as modal filtering, high-order cubature/quadrature integration, and geometric smoothing processes need to be introduced to suppress the poorly behaved components of the numerical schemes. However, these techniques are fully implemented in the codes accompanying this book and do not need to be rewritten from scratch. It is thus up to the reader to take what is written, explore with it, and apply it to the problems that interest them! What we are effectively telling the reader is that while mid-to-high order methods in general geometries may appear difficult or cumbersome, they are neither to a prohibitive degree. If the reader is seeking a high-resolution, geometrically flexible method, they need only follow the lessons explained herein.

References

[1] Frédéric Dupont, Comparison of Numerical Methods for Modelling Ocean Circulation in Basins With Irregular Coasts, McGill University, 2001.

[2] C. Geuzaine, J.-F. Remacle, A three-dimensional finite element mesh generator with built-in pre- and post-processing facilities, International Journal for Numerical Methods in Engineering 79 (2009) 1309–1331.

[3] W.N. Gordon, C.A. Hall, Construction of curvilinear coordinate systems and application to mesh generation, International Journal for Numerical Methods in Engineering 7 (1973) 461–477.

[4] J. Hesthaven, T. Warburton, Nodal Discontinuous Galerkin Methods, Springer, 2008.

[5] P.K. Kundu, I.M. Cohen, Fluid Mechanics, 4th ed., Elsevier Academic Press, 2008.

[6] S. Wandzura, H. Xiao, Symmetric quadrature rules on a triangle, Computers & Mathematics with Applications 45 (2003) 1829–1840.

[7] X. Zhang, Y. Xia, C.-W. Shu, Maximum-principle-satisfying and positivity-preserving high order discontinuous Galerkin schemes for conservation laws on triangular meshes, Journal of Scientific Computing 50 (1) (2012) 29–62.

CHAPTER

Beyond standard treatments: Flow in porous media

12

CONTENTS

12.1 Motivation... 225
12.2 The basic theory of porous media 225
12.3 Flow in saturated porous media: two simple examples 228
12.4 Flow in saturated porous media: a more complex example 231
12.5 Towards more general descriptions: unsaturated flow 233
12.6 Mini-projects 236
12.7 Concluding remarks 236
References... 237

12.1 Motivation

The previous three chapters have departed somewhat from the first principles modeling advocated for in Chapters 1–7. Both finite volume and finite element methods are mature subjects, with a voluminous literature and specialized notation. In the case of finite volume methods, they are also the methodology of choice for existing lake and coastal ocean models. While, we have made a solid argument for geometrically flexible, mid-order methods, we would like to conclude the book by returning to modeling.

Up to now, we have focused on fluid dynamics and tracers (i.e. the ecology) in natural bodies of water governed by the incompressible Navier-Stokes equations, especially in their simplified forms in the x-z and x-y planes, there is still much to be said about flow through the bottom sediments of these bodies of water, to the groundwater that lays underneath.

12.2 The basic theory of porous media

The basic theory of porous media presumes that there is some fraction of the medium occupied by a solid material (e.g., sand, silt, clay, or a combination thereof) and a remainder that may be occupied by the fluid in either gaseous or liquid phase. The fluid is said to be occupying the "pore space," or simply the pores. The fraction occupied by the fluid is referred to as the **pore fraction** and is denoted by φ (note the

Physics and Ecology in Fluids. https://doi.org/10.1016/B978-0-32-391244-0.00022-X
Copyright © 2023 Elsevier Inc. All rights reserved.

225

possible confusion with the vertical structure of linear internal wave theory discussed in Chapter 8). The pore space is typically thought of as being made up of very thin, tortuous passages with a schematic shown in panel (a) of Fig. 12.1. The mathematical problem of solving for the flow in the passages would begin with the Navier-Stokes equations, for which the momentum equation reads

$$\frac{D\vec{u}}{Dt} = -\nabla p + \rho\vec{g} + \mu\nabla^2\vec{u}$$

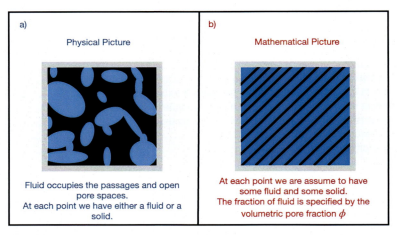

FIGURE 12.1

Physical picture vs. mathematical picture of porous media description.

Since the passages are typically very small in diameter and the flow through them is slow, we may neglect the acceleration term. Discarding gravitational effects for the moment and exploring only the so-called creeping-flow limit, we find

$$\mu\nabla^2\vec{u} = -\nabla p. \tag{12.1}$$

This is a linear equation and hence one may think it would pose few difficulties. However, since the microscopic geometry of the pore-space is essentially impossible to specify exactly we are left with a highly uncertain problem ([9]).

To avoid the problem of specifying the geometry of the pore space modelers have made a key mathematical assumption, that of a co-occupied continuum. This is schematized in panel (b) of Fig. 12.1. The physical porous material is averaged so that at each point there is a fluid and solid component with φ providing the ratio occupied by fluid at each point. All that remains is to specify the flow rate. While there are rigorous mathematical approaches to the averaging procedure for simplified configurations (see for example the book [6]), for the vast majority of situations the averaging is the invisible scaffold behind the theory and modeling simply proceeds from the final set of equations (which we show below).

12.2 The basic theory of porous media

To make progress we assume that the pore space is completely saturated by fluid the viscous frictional resistance is a linear function of the velocity times the unknown fraction of fill-able pore space φ, leading us to write

$$\varphi \vec{u} = -\frac{k}{\mu} \nabla p \,, \tag{12.2}$$

where $\underline{k}$ is referred to as the medium-dependent second-order **permeability** tensor (units of length), and $0 < \varphi < 1$ (dimensionless) is called the porosity of the medium and is left as a parameter. While φ is typically approximated as a constant it is, more generally, a function of the compressibility of the solid phase and increases (decreases) with the water pressure above (below) some equilibrium value ([1]). μ is the fluid viscosity.

Eq. (12.2) is known as **Darcy's Law** and forms the basis of the momentum balance for the majority of past and present flow theory in saturated porous media (see [10] for a more complete discussion). To gain further insight, we assume that the fluid is **incompressible**, as done throughout this book, i.e.,

$$\nabla \cdot \vec{u} = 0 \,.$$

After rearranging Darcy's law and taking the divergence we find that

$$0 = \nabla \cdot \vec{u} = \nabla \cdot (K \nabla p) \,, \tag{12.3}$$

where $K = k \mu^{-1} \varphi^{-1}$ is a parameter grouping and in general, represents a spatially-varying material coefficient. From here, one can recover a standard Poisson-type problem by specifying a geometric region of interest. When the various parameters are assumed constant one recovers Laplace's equation

$$\nabla^2 p = 0. \tag{12.4}$$

The problem formulation is completed by prescribing suitable pressure values (Dirichlet conditions) along the boundary, if known, or the no-flow condition at impermeable boundaries $\vec{u} \cdot \hat{n} = 0$ which is equivalent to

$$\frac{\partial p}{\partial n} = 0 \quad \text{along} \quad \partial \Omega \,,$$

for the case of Darcy's law. The reader has likely already noticed that unlike the Navier-Stokes equations, the above equation has no time derivative term. The system thus adjusts to time varying boundary conditions instantly. Moreover, the above equations imply that if the pressure is spatially independent along the boundary, the flow is identically zero. We will work out some simple problems in the next section. Here it is worth noting that while the Darcy's law based modeling has a long history in hydrology, from a modeling point of view it leaves much to be desired at first glance, since so much of the result is controlled by the simplicity of Darcy's law.

12.3 Flow in saturated porous media: two simple examples

In panel (a) of Fig. 12.2 we show a simple configuration of a lake with a porous bottom. In an actual lake the near bottom region is quite complex, with biological detritus mixing with near bottom mud to create a layer that is better described as a non-Newtonian fluid. Nevertheless, to keep the discussion focused we consider a clear water layer overlying a porous medium (for concreteness, think of an oligotrophic lake with a sandy bottom).

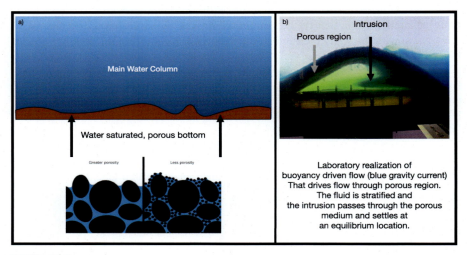

FIGURE 12.2

(a) Porous lake bottom conceptual diagram and **(b)** laboratory demonstration photo of buoyancy driven flow over a porous bottom driving flow through the porous region.

To illustrate to the reader that motions in the water column can drive non-trivial motions through a porous medium, panel (b) of Fig. 12.2 shows a laboratory experiment in which a density driven flow moves over a porous hill. The flow is an intrusive gravity current propagating near the interface between two different densities of fluid (the denser fluid is dyed green, and the intruding fluid is dyed blue). The complex spatio-temporal pattern of pressure at the interface of water column and the porous medium drives flow through the porous medium (which is impossible to see). However, this flow is manifested as a spreading layer of dark fluid in the open fluid beneath the porous medium.

The open region filled with fluid beneath the porous medium is a laboratory tool used to visualize the fluid motion, and hence is not representative of what goes on in an actual lake. Nevertheless, it illustrates that fluid motions above the bottom sediment can drive significant motion in the porous medium. This has been observed in the literature, for example [7].

12.3 Flow in saturated porous media: two simple examples   **229**

Let us see if the laboratory observations are easy to replicate in the mathematical model. First consider a trivial case in which pressure at the interface is time dependent

$$p(z = 0, t) = f(t)$$

for some known function $f(.)$. Assume the flow is one dimensional with the porous layer overlying an impermeable layer at $z = -D$ (think of sand over a layer of clay). The condition of no flow at $z = -D$ gives (from Darcy's law)

$$\frac{dp}{dz}|_{z=-D} = 0.$$

The governing equation reduces to an ordinary differential equation

$$\frac{d}{dz}\left(K(z)\frac{dp}{dz}\right) = 0$$

where we have not assumed the various parameters are constants. This equation may immediately be integrated once to yield

$$K(z)\frac{dp}{dz} = c_1(t)$$

where $c_1(t)$ is constant of integration (which may have a parametric dependence on time). It turns out that what is important is the boundary condition at the bottom of the porous layer, which specifies no flow. This gives

$$c_1(t) = 0$$

and canceling $K(z)$ and integrating gives

$$p(z, t) = c_2(t).$$

Substituting the boundary condition at $z = 0$ gives

$$p(z, t) = f(t). \tag{12.5}$$

The pressure is thus completely specified by the pressure at the fluid-porous medium interface and there is no flow at any point in the domain.

While somewhat useless, this is an important solution to have in one's arsenal, because it illustrates the rather strong assumptions the incompressible Darcy based theory makes.

Darcy's law shines a spotlight on spatial gradients of pressure. For our second problem we return to the linearized seiches in a lake with a flat bottom and ask whether this is enough to drive a non-trivial flow in the porous medium beneath.

For concreteness return to the porous layer of thickness D with an impermeable layer at $z = -D$. To drive the flow in the porous medium consider an "aquarium"

230 **CHAPTER 12** Beyond standard treatments: Flow in porous media

lake of length L with a constant depth H. The seiche will be given by the solution of the linearized shallow water equations (hence we can safely ignore any dispersive effects) so that the free-surface is given by

$$\eta(x, t) = \eta_0 \cos(\sigma t) \cos(\pi x / L).$$

If we neglect atmospheric pressure (a safe assumption due to the high density of water) the pressure at the bottom is given by

$$p = \rho_0 g (H + \eta).$$

For simplicity assume that the various parameters that characterize the porous layer are constants so that the pressure in the porous medium is given by Laplace's equation (12.4). Laplace's equation is linear so that the only piece we need to consider for driving the flow in the porous medium is

$$p(x, z = 0) = \eta_0 \rho_0 g \cos(\sigma t) \cos(\pi x / L)$$

and indeed even the constant out front may be ignored. Since we know that differentiating a cosine twice gives us another cosine we make the inspired guess of a solution:

$$p = p_f \cos(\sigma t) \cos(\pi x / L) \hat{g}(z).$$

The hat symbol is used to differentiate the function $\hat{g}(z)$ from the acceleration due to gravity g. Notice the way the time dependence again only appears as a parametric factor and does not influence the form of the spatial distribution of the solution in the porous medium. A quick bit of algebra gives

$$\frac{d^2 \hat{g}}{dz} - \frac{\pi^2}{L^2} \hat{g} = 0$$

with the boundary conditions $\hat{g}(0) = 1$ and $\hat{g}'(-D) = 0$. We leave the calculation for p_f to the reader as an *Exercise*. The solution is a textbook problem in second order differential equations and gives

$$\hat{g}(z) = c_1 \exp(\pi z / L) + c_2 \exp(-\pi z / L).$$

We can however, be a little more tricky and rewrite the solution so that the condition at the impermeable layer is satisfied automatically. We do this by using the hyperbolic cosine, or

$$\cosh(A) = \frac{\exp(A) + \exp(-A)}{2}$$

so that the solution is written

$$\hat{g}(z) = M \cosh \left(\frac{\pi}{L} (z + D) \right) \tag{12.6}$$

12.4 Flow in saturated porous media: a more complex example 231

where M is the constant of integration that will allow us to satisfy the boundary condition at the fluid-porous medium interface. A bit of algebra gives

$$\hat{g}'(z) = \frac{M\pi}{L} \sinh\left(\frac{\pi}{L}(z+D)\right) \tag{12.7}$$

so that $\hat{g}'(-D) = 0$ as we wished. At the fluid-porous medium interface we have $\hat{g}(0) = 1$ so that

$$M = \frac{1}{\cosh\left(\frac{\pi D}{L}\right)}.$$

The flow through the fluids-porous media interface is proportional to $\frac{\partial p}{\partial z}$ and hence it is $\hat{g}'(z)$ that is of interest for transport calculations. From (12.7) we see that this function is inversely proportional to the lake length. This suggests that flow due to lake scale fluctuations like seiches is not that important, at least within the scope of the present version of the theory (see Mini-project 1 for further discussion).

It is interesting that we were able to conclude this without resorting to numerical methods. This is completely due to the linear nature of the incompressible, Darcy's law based theory. Indeed, the reduction to a Laplace's equation when the parameters characterizing the porous medium are constant allows one to use hundreds of years of mathematical research on the Laplace equation.

12.4 Flow in saturated porous media: a more complex example

In the previous section, we mentioned that when the parameters that characterize the porous medium are constants, the problem for the pressure reduces to Laplace's equation. This is perhaps the most well studied equation in applied mathematics, and in this section we will make use of this fact.

The goal, as far as a physical problem is concerned, is to have a solution for the flow induced over a small scale feature. Recall that in the previous section we showed that lake scale seiches are not really expected to yield strong flow across the fluid-porous medium interface. This was due to a factor of π/L in the solution of the velocity.

Solutions of partial differential equations, when available as closed form formulae, are typically quite complex. This means that we cannot really expect solutions for any kind of bottom bump. However, a bump in the shape of half circle is about as symmetric as we can get, and this is the configuration we will choose. Fig. 12.3 shows a diagram of the situation. Making the simplest possible, non-trivial choice, we assume the flow far from the bump is time independent and uniform.

232 CHAPTER 12 Beyond standard treatments: Flow in porous media

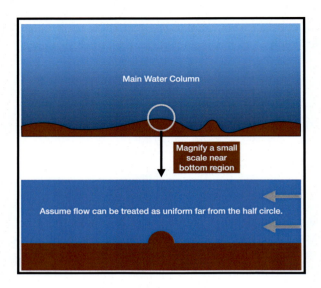

FIGURE 12.3

Porous lake bottom with a region magnified to show a small, semi-circular bump on the porous bottom. The flow is assumed to be uni-directional and uniform far from the bump. In the text we explain how this can facilitate an exact solution.

We now have two problems to solve. One is physical: we need some sensible way to get the pressure at the boundary. The other is purely mathematical: we need to solve Laplace's equation in the half circle.

To address the first problem, we take an extremely utilitarian approach. In the theory of fluid mechanics ([8]) potential flow is one of the oldest sub-disciplines, and while potential flow makes some relatively strong assumptions (no near bottom boundary layers for one) it is a good place to start for exact solutions. There is a potential flow solution for flow past a circle, and it turns out to be symmetric across the symmetry line of the circle that coincides with the current (in our example this is the flat bottom). We can thus borrow this solution. The only catch is that the solution is given in polar coordinates with the radial coordinate scaled by the radius of the semi-circular bump. The formula form of the solution reads

$$p(1, \theta) = 2\cos(2\theta) - 1. \qquad (12.8)$$

This means the pressure is highest at front and back of the bump, where the bump meets the flat porous bottom, and lowest at the top of the bump. Even without doing any calculations we thus expect from the nature of Darcy's law to get flow out of the top of the bump!

The incompressible Darcy theory can be solved in the semi-circle (a far more general discussion of this is available in [12]) to give

$$p = 2r^2 \cos(2\theta) - 1. \tag{12.9}$$

According to Darcy's law, the radial flow is proportional to the radial component of the gradient of pressure, and this simplifies to

$$\varphi u^{(r)} = -\frac{4kr}{\mu} \cos(2\theta) \tag{12.10}$$

and this formula confirms the intuitive idea from above: there is flow from the fluid into the porous medium near the front and back of the semi-circular obstacle, and a compensating flow from the porous medium into the open water column near the top of the semi-circular bump. This could have practical importance for oligotrophic lakes, for which the interior of the water column is essentially a desert. Since dead organisms sink to the bottom, the resulting source of nutrients may well end up buried in sediments. Any mechanism that systematically transports material across the fluid-porous medium interface is very much worth careful study.

The theory developed above finds some support in experiments ([7]), though the assumption of potential flow is overly simplistic. Indeed the fact that flow along a porous material induces some flow inside the porous medium has been known for decades, with a classical (though often criticized) model presented in [2]. The interested reader is encouraged to follow the active literature in this field, and formulate some simple models for themselves.

There is one more modeling point worth making. The classical theory of porous media assumes that the porous medium is akin to a sponge (or a piece of swiss cheese if your imagination does better with food based analogies). Of course, an actual lake bottom is nothing like this. In the bottom diagram in panel (a) of Fig. 12.2 we schematize an interesting possibility. Here the porous medium consists of the classical pore space, but portions of this pore space are filled by smaller particles. If fluid motions external to the porous medium efficiently pump fluid out of the pore space, they would in turn bring out some of the smaller particles. This would in turn change the porosity and permeability, and would allow for an interesting feedback mechanism between external flows and flows within the porous medium. This idea was explored using internal waves as the external flow in [11], though its efficiency in the field has (to the best of our knowledge) not been measured.

12.5 Towards more general descriptions: unsaturated flow

While the theory of fluid mechanics, leading to the Navier-Stokes equations, is well-validated experimentally and has clear ties to first principles, the theory of saturated flow in porous media is somewhat murkier. To finish our narrative, we briefly discuss

234 CHAPTER 12 Beyond standard treatments: Flow in porous media

an even more complex situation, that of unsaturated porous media theory, for which lively arguments about the basic equations continue to persist.

For this section we do not assume the porous medium is saturated with water, as one might find unsaturated sands and soils near the coastline of a natural body water, nor do we assume incompressibility of water within the medium. In such cases, we might turn to the basic continuity equation (or 'conservation of mass')

$$\frac{\partial \rho}{\partial t} + \nabla \cdot (\rho \vec{u}) = 0 \, .$$

Notice that substituting Darcy's law into the above would result in a time-dependent equation. However, for a general description that allows only for water to occupy the pore space that is in general unsaturated, we must make the substitutions

$$\rho \leftarrow \varphi S \rho \, , \, \rho \vec{u} \leftarrow \rho \varphi \vec{u} \, ,$$

where $0 < S \leq 1$ is the fraction of the pore-space filled with water, or the *saturation*. If we presume the solid-phase does not move with the water phase and that air density is a constant, the mass balance equation reads,

$$\frac{\partial (\varphi S \rho)}{\partial t} + \nabla \cdot (\rho \varphi \vec{u}) = 0 \, . \tag{12.11}$$

Writing down Darcy's law with the density-dependent gravity term included, we find

$$\varphi \vec{u} = -\frac{k_r K_A}{\mu} (\nabla p + \rho \vec{g}) \, . \tag{12.12}$$

We now have two equations (mass balance and momentum balance) and three unknowns. We have also expanded the permeability k to include a relative permeability k_r and absolute (time-independent) permeability K_A, where in general, the relative permeability is a function of the capillary pressure P_{caw} in unsaturated flow. To make progress, it is typically assumed that

$$P_{caw} = p_{air} - p \, ,$$

where p is the usual water pressure and p_{air} is the air pressure, assumed constant. The subscript "caw" stands for "capillary air water". Since P_{caw} is not known, it must be derived from thermodynamic considerations, and usually, it is assumed to be of the form $P_{caw} = P_{caw}(S, T)$ where T is temperature (usually assumed constant, also). Under laboratory conditions, it is possible to measure and calculate the pressure difference $p_{air} - p$ while varying the water content, allowing one to explore empirical relationships.

Further simplifications arise under the assumption of a constant density $\rho = \rho_0$, wherein we find the pair of equations

$$\frac{\partial \theta}{\partial t} + \nabla \cdot (\theta \vec{u}) = 0 \, , \tag{12.13}$$

12.5 Towards more general descriptions: unsaturated flow   **235**

$$\vec{q} = \theta\vec{u} = -\tilde{K}(\psi)\left[\nabla\psi + \vec{k}\right].\qquad(12.14)$$

Here $\theta = \varphi S$ is the *volumetric water content* (volume of water per volume of medium), $\tilde{K}$ is the *hydraulic conductivity* (units of speed), and $\psi = p/\rho_0 g$ is the *matric potential* or the matric *head*. The matric potential represents the attraction between water and the porous matrix. The word matric is thus meant in the sense of "relating to a matrix", where matrix is not a mathematical matrix, but the solid part of the porous medium. A total potential function, or total head can be defined as

$$h = \psi + z,$$

where we use the fact that $\nabla z = \vec{k} = (0, 0, 1)$. We finally write down a "single" equation governing unsaturated flow as

$$\frac{\partial\theta}{\partial t} = \nabla\cdot\left(\tilde{K}\nabla h\right).\qquad(12.15)$$

Of course, "single" is in quotation marks since additional constitutive laws such as Van Genuchten's [13] functions or the Brooks-Corey [3] relations are used to close off the system as $\tilde{K} = \tilde{K}(\psi) = k_r(\psi)K_A g\rho_0/\mu$ is a function of the matric potential ψ, and it is generally assumed that ψ's dependence on the dependent variables is furnished via its dependence on θ so that, $\psi = \psi(\theta)$ only, as a consequence of the relation for P_{caw} under the assumption of constant air pressure, where $\theta = \theta(x, y, z, t)$.

Eq. (12.15) is known in the hydrology literature as the "Richards equation." While the Richards equation with appropriate constitutive laws represents the current "state of the art" in unsaturated flow modeling, it has several pitfalls. For instance, its material coefficient has a non-linear dependence on the unknown dependent variable ψ, and this in turn depends on the unknown moisture content variable θ. As a result of this nonlinear coupling, the numerical solution is not, in general, well-behaved from a numerical analysis (i.e., stability) perspective unless $\tilde{K}(\psi)$ and $\psi(\theta)$ are smooth and monotonic single-variable functions.

There are also several known theoretical pitfalls. As one example, we can consider what happens in the case of a hydrostatically balanced state when $\vec{u} = 0$. In such situations, we immediately recover a matric potential head that is a linear function of the elevation coordinate z, which is in direct contradiction with the equation of state for $P_{caw}(S, T)$ which must be a function of the saturation, and other material properties. The water pressure p is another area of high contention, since soil experimentalists have argued that it can attain negative absolute values, despite its thermodynamic definition which guarantees $p > 0$ everywhere.

More must be done to correctly develop a physically consistent theory of unsaturated flow before a truly general numerical treatment may be developed and implemented, as was possible with saturated flow above, or with general non-porous fluid flows throughout the rest of this book. This subject has been a topic of discussion in the literature, particularly in the critique by Gray and Hassanizadeh [5]. This paper proposes an alternate theory, and argues that a dependent variable representing the fractional quantity of air-water interface is of fundamental importance and

236 CHAPTER 12 Beyond standard treatments: Flow in porous media

hence must appear in the governing equations. The authors develop their theory by beginning with the equations for two-phase flow, and proceed to reduce the equations to an averaged macroscopic theory of unsaturated single-phase flow, arguing that the generic derivation of the Richards equation, as depicted above, is not systematic since it introduces the relations $k_r = k_r(\psi)$ and $\psi = \psi(\theta)$ in an *ad hoc* manner, so as to form a closed system of equations. The authors of this book agree that it is perhaps more sensible to start with known conservation laws and thermodynamic equations of state, and then reduce the system to a simpler theory, as opposed to be beginning with a simple theory (single-phase flow) and introducing new relations *ex post facto*. In any event, this section provides the reader with a clear example where theoretical work remains to be done by future research.

12.6 Mini-projects

1. Research the Beavers-Joseph boundary condition. How would it modify the problem of the flow in the porous bottom due to a seiche? Would we expect more induced flow? If so, where?
2. There are two well known extensions of Darcy's law. One is called the Brinkman extension and the other the Forchheimer extension. Using a suitable reference (e.g. [10]) discuss the mathematical form of, and physical reasoning behind, each.
3. There are formulations of porous media theory that allow for a smooth transition between the open water column and the porous medium. One flavor goes by the name "Brinkman penalization theory". Using an appropriate source explain how these generalize the Navier-Stokes and Darcy equations.
4. The fact that the incompressible Darcy theory reduces to a Laplace's equation or its generalization has been exploited in a variety of semi-analytical methods. One of these is given in [4]. Explain how existing mathematics is used in this method, and comment on how the method would be implemented in practice.

12.7 Concluding remarks

With this chapter, we have concluded our short tour of mathematical and computational modeling of flow in natural waters and its effect on active tracers (effectively population biology). We have made no claim as to completeness; modern scientific effort is broad and very far from unified. Indeed, as a quick wander through the poster session of a large international conference (e.g. the annual AGU Fall Meeting) shows, human ingenuity is truly amazing.

The availability of inexpensive computation and various kinds of mathematical software (whether completely open source, or partially open due to subsidies by academic institutions for their members) is a game changer as far as what one can accomplish locally. However, systematic courses that teach students how to experiment, critique their experiments, and produce meaningful scientific graphics are few and far between. This is complicated by the fact that not everyone has the same set of tastes when it comes to research.

We have tried to offer multiple introductory paths to the reader. Some may be laser focused on basin scale modeling and for these readers we hope to have provided a clear critique of low order models and their exceedingly high numerical diffusion. For readers interested in biogeochemistry we hope to have provided a sense of the interplay with dynamical systems theory, that may be lost when more complex models are employed. We have also provided the step from ordinary differential equation models to partial differential equation models, through the mechanisms of advection and diffusion. Reaction diffusion is a standard topic in mathematical biology, but advection is often overlooked. This is not possible in natural waters, and we have provided plenty of hands-on evidence for this. Finally, we have presented individual based models. These form a completely different class of theoretical and numerical model, and in our opinion have considerable upside in many fields (some well outside the scope of this book).

The codes available online are a vital part of this book. Indeed, it is safe to say that both authors read as much code as words. But books and journal articles are a very different social construct from software. Software is meant to be run, and reading code is either an indication that something has gone wrong (i.e. the process of debugging) or that the code has been run and the user now wants to see the inside of the machine. We have thus sought to give the reader the chance to actively explore, using codes that run effectively for us. Once this exploration has gone on for some time, perhaps a user may even decide that they can write much better code than we have. We think this is great, and hope that they share this code with the broader community. A user may also feel like it is time to use a more complex model in the public domain. Some of these models have active, open user communities, e.g. ROMS

```
https://www.myroms.org/
```

or the MITgcm

```
https://mitgcm.readthedocs.io/en/latest/
```

They are, by construction, very different tools to the codes included with this book, but they may be more suitable to a reader's long term research needs. The role of the small code is thus to provide the foundation on which to build explorations with the more complex code.

There is a lot of fertile ground for research, and there is no risk of running out of interesting problems. As great as modern science is, a complete quantitative model of the hydrodynamics and ecosystem in a local reservoir is some way off into the future.

References

[1] B. Aberg, Void ratio of noncohesive soils and similar materials, Journal of Geotechnical Engineering 118 (9) (1992) 1315–1334, https://doi.org/10.1061/(ASCE)0733-

9410(1992)118:9(1315), https://ascelibrary.org/doi/abs/10.1061/%28ASCE%290733-9410%281992%29118%3A9%281315%29.

[2] Gordon S. Beavers, Daniel D. Joseph, Boundary conditions at a naturally permeable wall, Journal of Fluid Mechanics 30 (1) (1967) 197–207.

[3] R.H. Brooks, A.T. Corey, Hydraulic properties of porous media, Computers & Geosciences 3 (1964), https://doi.org/10.1016/j.cageo.2021.104897.

[4] J.R. Craig, A general analytical solution for steady flow in heterogeneous porous media, Water Resources Research 51 (6) (2015) 4184–4197.

[5] William G. Gray, S. Majid Hassanizadeh, Paradoxes and realities in unsaturated flow theory, Water Resources Research 27 (8) (1991) 1847–1854, https://doi.org/10.1029/91WR01259, https://agupubs.onlinelibrary.wiley.com/doi/abs/10.1029/91WR01259.

[6] Ulrich Hornung, Homogenization and Porous Media, vol. 6, Springer, 1996.

[7] Markus Huettel, Wiebke Ziebis, Stefan Forster, Flow-induced uptake of particulate matter in permeable sediments, Limnology and Oceanography 41 (2) (1996) 309–322.

[8] P.K. Kundu, I.M. Cohen, Fluid Mechanics, 4th ed., Elsevier Academic Press, 2008.

[9] Robert S. Maier, et al., Pore-scale simulation of dispersion, Physics of Fluids 12 (8) (2000) 2065–2079.

[10] Donald A. Nield, Adrian Bejan, Convection in Porous Media, vol. 3, Springer, 2006.

[11] Jason Olsthoorn, Marek Stastna, Nancy Soontiens, Fluid circulation and seepage in lake sediment due to propagating and trapped internal waves, Water Resources Research 48 (11) (2012).

[12] Jared Penney, Marek Stastna, Numerical simulations of flow through a variable permeability circular cylinder, Physics of Fluids 33 (11) (2021) 117113.

[13] M.Th. van Genuchten, A closed-form equation for predicting the hydraulic conductivity of unsaturated soils, Soil Science Society of America Journal 44 (5) (1980) 892–898, https://doi.org/10.2136/sssaj1980.03615995004400050002x, https://acsess.onlinelibrary.wiley.com/doi/abs/10.2136/sssaj1980.03615995004400050002x.

Glossary

Advective time scale The ratio of a typical length scale and typical velocity scale. Effectively the time taken for a fluid parcel to travel a fixed distance.

Aspect ratio The ratio (or fraction) of a typical horizontal length scale to a typical vertical (depth or height) scale.

Buoyancy frequency The frequency of a oscillation of vertically displaced fluid parcel in a stably stratified fluid. Proportional to the square root of the vertical derivative of the density stratification function. Sometimes referred to as the Brunt Väisälä frequency.

Carrying capacity The maximum population that can be achieved given the limitations of the surrounding environment.

Conservation Law (mathematical) A system of hyperbolic equations written as a combination of temporal derivatives, the row-wise divergence of a matrix function, and source terms. Vital for many finite volume and DG methods.

Conservation Law (physical) The governing partial differential equation for the evolution of a physical quantity (e.g. mass, linear momentum, energy).

Coriolis Effect The effect of a rotating system, like the Earth, that appears to deflect motion to the right when viewed in the rotating frame.

Darcy's Law Specifies the flow in the pore space as proportional to the imposed pressure gradient. While phenomenological, as opposed to first principles, it is the fundamental law of flow in porous media.

Density Stratification The naturally occurring layering of waters with less dense (i.e. warm) water overlying more dense (i.e. cold) water.

Discontinuous Galerkin (DG) Method A class of finite element method which does not impose continuity at element boundaries.

Dispersion The property of most naturally occurring waves which gives wave speeds (both phase and group) that change with wavelength.

Dispersive Correction A correction to the assumption of purely hydrostatic pressure in the context of the shallow water equations.

Dispersive Waves Waves that exhibit dispersion.

Eddy viscosity The parameter measuring the effective diffusion of momentum by unresolved, turbulent motions. Typically orders of magnitude larger than molecular viscosity. Assumes unresolved scales behave like diffusion.

Entropy A quantitative measure of the amount of "disorder" in a closed system. In terms of elementary physical quantities, it is given by the amount of reversible heat divided by the temperature.

f-plane The commonly used approximation for representing the Earth's rotation on the scale of a lake or reservoir. It leads to a local Cartesian set of axes and a new horizontal pseudo-force in the horizontal momentum balance on these axes.

Fast Fourier Transform (FFT) A "fast" algorithm, usually attributed to Gauss, for computing the coefficients of the sinusoidal components in the decomposition of a signal.

Finite Element Method A class of numerical method based on local approximations in individual "elements" that are combined into a description in the whole domain.

Finite Volume Method A class of numerical method based on the integral form of the conservation law that leads to the PDE being discretized.

240   Glossary

Frequency Given as the inverse of the wave period (multiplied by 2π). Specifies how quickly waves oscillate temporally.

Froude number A dimensionless parameter which measures the relative strength of the buoyancy and inertia terms. Low Froude number implies buoyancy is more important. Usually specified as a ratio of velocity and a typical wave speed in the system.

Geostrophic balance A reduced form of the fluid mechanical momentum equations in which the rotation terms are balanced by the gradient of pressure.

Gravity Wave Wave motion for which the restoring force is given by gravity.

Group Velocity The velocity with which energy propagates for dispersive waves.

Hovmoeller Plot A type of contour plot with a spatial and temporal axes. Often used to demonstrate the presence of waves. Sometimes referred to as a space-time plot.

Hydrostatic Pressure The component of pressure given by the weight of the overlying fluid. The dominant part of pressure for large scale flow.

Internal Gravity Wave Wave motion that occurs in the interior of natural waters due to fluctuations about the naturally occurring density stratification.

Internal Seiche A periodic motion in the interior of a stratified, closed body of water.

Kelvin Wave A wave that is largest at the coast and decays exponentially away from the coast. Possible only due to rotation AND in the presence of boundaries.

Large Eddy Simulation (LES) A decomposition of fluid variables on a discrete grid into a resolved and unresolved part. A numerical modeling alternative to RANS.

Material derivative The rate of change of a quantity following a fluid particle. It includes a term that expresses the rate of change in time, and a group of terms due to the flow of the fluid.

Navier-Stokes Equations The governing equations for a viscous, Newtonian fluid.

Newton's Second Law Sir Isaac Newton's law of motion that states that the time rate of change of linear momentum of a body is equal to the sum of the forces acting on the body. Often abbreviated in an overly simplistic manner as "$F = ma$."

Pathline The path traced by a fluid particle in a fluid flow. Almost never available as a formula.

Phase Velocity The velocity with which wave crests propagate for dispersive waves.

Planktor An individual planktonic organism (could be either phyto or zoo plankton).

Poincaré Waves Rotation modified surface gravity waves without boundaries.

Population The total number of members in a species. Though in practice this quantity is a non-negative integer, it is often represented as a non-negative real number in mathematical models.

Position A vector (given by e.g., $\vec{x} = (x, y, z)$ in three dimensions) denoting a location relative to the origin (at e.g., $(0, 0, 0)$).

Power Spectral Density The square of the complex modulus of the Fourier transform of a time-signal of interest, i.e., a time series. Illustrative of the relative importance of the various sinusoidal (frequency-domain) components of the signal.

Relaxation oscillation An oscillation marked by a sharp, fast increase followed by a slower, gradual decrease.

Reynolds Averaged Navier-Stokes (RANS) The name given to the equations of fluid motion once a turbulent decomposition and ensemble averaging have been applied.

Reynolds Averaging The mathematical process of splitting a physical variable into an ensemble mean and a fluctuation about this mean.

Reynolds Number A dimensionless number that measures the relative importance of the inertia and viscous terms in the governing equations. Large Reynolds number (that is typical of natural waters) implies viscosity is less important.

Reynolds Stress The new term appearing in the equations of motion upon applying a turbulent (or Reynolds) decomposition and ensemble averaging the momentum equations. It expresses the change in mean momentum induced by fluctuations.

Rossby Number A dimensionless number that measures the relative importance of the inertia and rotation terms in the governing equations. Low Rossby number (that is typical of large lakes) implies rotation is more important.

Second Law of Thermodynamics, the A fundamental law of nature stating that the *entropy* (see definition above) of a closed system must not decrease, unless the system undergoes work in response to external forcing. As a consequence of the "second law", the mechanical energy of a closed system must be non-increasing: $\frac{dE}{dt} \leq 0$.

Seiche A periodic motion in a closed body of water. Sometimes referred to as the bathtub, or sloshing mode. This simple characterization is modified by rotation.

Shallow Water Equations Simplified equations for hydrostatic flow with a free surface.

Smagorinsky model A well known, if simplistic, model of the large-eddy simulation (LES) sub-grid-scale stress τ^{LES}. Here, τ^{LES} is taken to be the product of a modeled variable eddy-viscosity and the resolved rate of strain tensor, where the variable eddy-viscosity is itself a function of the resolved rate of strain tensor.

Spectra Plural of 'spectrum' – a set of discrete or continuous properties associated with a signal. In the case of the "Fourier spectrum", it consists of the sinusoidal coefficients of the various oscillatory frequencies that make up the signal when summed together.

Spectral Method A class of numerical methods for which the order of accuracy increases as the number of grid points increases. Often also called a Pseudospectral Method.

Stable Limit Cycle An orbit within a phase portrait that exhibits oscillatory behavior. Paths away from this orbit are attracted to it.

Streamfunction A function whose level curves yield streamlines.

Streamline Tangent curve to the velocity field at any fixed time. For a general flow field streamlines do not coincide with pathlines.

Surface Gravity Wave Wave motion that occurs at the interface of the atmosphere and a natural body of water.

Turbulence The irregular, chaotic, vortex rich flow of fluids, usually at high Reynolds number. Typical of natural waters.

Turbulence Closure An *ad hoc* fix for the Turbulence Closure Problem. Often involves a form of eddy viscosity and an assumption that turbulence acts analogously to molecular diffusion.

Turbulence Closure Problem A term expressing the fact that ensemble averaging the equations of motions does not lead to a closed set of equations. The problem cannot be solved by deriving equations for higher moments.

Viscosity The physical parameter specifying the rate of diffusion of linear momentum. Colloquially, the "stickiness" of a fluid.

Vorticity The local rotation of a fluid particle, expressed as $\nabla \times \vec{u}$.

Wavenumber Given as the inverse of the wave length (multiplied by 2π). Specifies how quickly waves oscillate spatially.

Wavetrain A modulated collection of individual waves. Applied to finite amplitude, or non-linear waves.

Index

A

Acceleration, 53, 54, 93, 226
Adjustment problem, 95, 96
Advection, 1, 6, 31–33, 37
    diffusion population model, 46
    linear, 207
    terms, 37
Advection-diffusion, 39
Advection-reaction-diffusion model, 39
Advective
    term, 53
    time scale, 67
Adverse pressure gradient, 213, 220
Aggregation, 156
Aliasing, 199, 210
Approach
    FEM, 184
    FV, 183
    Galerkin, 191
    nodal, 195
    scalar, 205, 206
Aspect ratio, 56, 57
Atmospheric pressure, 59, 230
Auxiliary variable, 198
Averaging
    kernel, 120
    Reynolds, 115, 120, 121, 130

B

Background flow, 169
Baroclinic speed, 77
Barotropic
    instability, 89
    seiches, 104
    speed, 77
Basic models, 3, 9
Basin, 66
    circular, 83, 219
Bathymetry, 197
Bed slope, 189
Bernoulli's equation, 213
Bessel equation, 80
Bessel function, 81
Biological tracers, 82, 165
Biophysical parameters, 33
Birth rate, 14, 15, 19, 20, 22–24
Birth-death model, 156
Blend factor, 195

Blending, 215
    function, 215
Body force, 55
Boussinesq approximation, 133
Box car filter, 120
Brinkman extension, 236
Brinkman penalization theory, 236
Broad band response, 24
Brunt-Vaisala frequency, 136
Buoyancy frequency, 136, 137
Burgers
    equation, 36, 115, 116, 120, 121, 123
    Johannes, 36
Bursting, 26

C

Carrying capacity, 14, 26, 35, 36
Channel, 66
Chaos, 24
Characteristic, 202
    curves, 202
    polynomial, 70
Circular
    basin, 83, 219
    domain, 66, 89, 214
    lake, 66, 79, 81
Circulation, 62, 64
Closed form, 36, 52
    solution, 31, 33
Co-occupied continuum, 226
Coastline, 71, 220, 234
Coefficient
    constant, 23
Cole-Hopf transformation, 36
Collocation method, 191
Collocation points, 191
Computational fluid dynamics, 149
Computing spectra, 28
Concentration, 32, 36
Condition
    boundary
        dynamic, 59
        kinematic, 58
        periodic, 37
    entropy, 170–172
    initial, 18
Conservation
    discrete, 168

243

244 Index

form, 59, 60
law, 188, 202
Constitutive law, 56
Continuum, 54
force, 55
Hypothesis, 51
Convective term, 53
Convergence speed, 183
Coriolis parameter, 221
Coriolis pseudo-force, 189
Cubature, 210, 217, 223
Curvilinear, 188, 189, 195, 198, 209, 210
elements, 214, 219
Cut-cells, 184

D

Darcy's law, 227, 229
Death rate, 14, 15, 20, 23, 26
Decay, 13, 20, 23, 32, 33
rate of, 20
slow, 22
Decomposition, 126
Reynolds, 119
Density
fluid, 54
power spectral, 21, 24, 29
Deterministic distribution, 40
DG-FEM, 209
solutions, 210
Differential equation, 53
theory, 15, 23
Diffusion, 31, 33, 36–38
effective, turbulent, 117
equation, 33, 36
molecular, 117
numerical, 173
Dimensional splitting, 176
Discontinuous Galerkin Finite Element Method
(DG-FEM), 7, 86, 184, 187–190, 194,
197, 206, 210, 220
Discontinuous Galerkin methods, 184, 209
Discrete conservation, 168
Discretization, 166
Dispersion, 93, 124
numerical, 126
relation, 69, 70, 75
shear, 43
Taylor, 43
wave, 188, 189, 197
Dispersive correction, 93, 112, 130
Dispersive shallow water
equations, 188
model, 92

system, 181
Dispersive wave, 70, 183
Dissipation
numerical, 126, 128
Divergence, 125, 127, 129
Double jet, 66, 86, 88
Duffing oscillator, 28
Duffing's equation, 13
Dynamic boundary condition, 59
Dynamical systems, 26
theory, 36
Dynamics, 2
computational fluid, 149
stratified fluid, 133

E

Ecological, 19
Eddies, 210
spurious, 210, 213, 214, 219, 220
Eddy viscosity, 56, 115, 117–119, 122–124, 128
effective, 128
Eigenvalue, 15
analysis, 15
Elliptic problem, 181
Energy, 170
analysis, 170, 171
mechanical, 13
potential, 11
total, 11, 30
Ensemble average, 116
Entropy condition, 170–172
Equation
Bernoulli's, 213
Bessel, 80
Burgers, 36, 115, 116, 120, 121, 123
differential, 53
ordinary, 10, 23, 33, 54
partial, 32, 33, 69, 166
theory, 15, 23
diffusion, 33, 36
dispersive shallow water, 188
Duffing's, 13
Fisher-KPP, 36
Fisher's, 36, 37
KdV, 92
Laplace's, 211, 212, 227
linear advection, 166
logistic, 35
momentum, 59–61, 67, 71, 72, 93, 111, 179,
199, 206
Navier-Stokes, 56, 116
reaction
advection, 32

Index **245**

diffusion, 32
Richards, 235
shallow water, 57, 60, 64, 66, 92, 115, 124, 125
wave continuity, 199
Equilibrium, 15
stable, 14, 20, 22, 23
Ergodic theorems, 116
Eulerian, 52
transport, 139
variable, 139
Evolution, 37, 82, 86, 96, 98, 146, 147
population, 31
seiche, 102
tracer, 43
Exchange, 20, 22
rate, 20, 23
Exponential
decay, 33
growth, 13–15, 33, 35
law, 35
law, 32
model, 32
Extension
Brinkman, 236
Forchheimer, 236
External Rossby deformation radius, 76

F

F-plane, 115, 124, 125
Factor
blend, 195
warp, 195, 215
Fast Fourier Transform (FFT), 29, 33, 37, 110
FEM approach, 184
Field
velocity, 42, 43, 139, 211
Filtering parameter, 199
Finite element method (FEM), 184
Finite volume
method, 166, 188, 202
Godunov-type, 183, 188
scheme, 170
Fisher-KPP equation, 36
Fisher's equation, 36, 37
Flow, 43, 51, 211, 213
background, 169
geostrophic, 125
potential, 64, 211, 232, 233
saturated, 227
subcritical, 61
supercritical, 61
unsaturated, 233, 234

Fluid
density, 54
mechanics
models, 26
Newtonian, 56
parcel, 52
pressure, 54
viscosity, 227
Flux
differencing, 168
Jacobian matrix, 179
Lax-Friedrichs, 185, 190, 205
numerical, 188, 190, 205
upwind, 169
upwind-biased, 171
Food web, 32
Forchheimer extension, 236
Form
conservation, 59, 60
quasilinear, 178
Fourier
analysis, 30
series, 28
space, 33, 121
spectral method, 33
transform, 33, 37
Frequency, 70
Brunt-Vaisala, 136
buoyancy, 136, 137
Front
propagating, 36
reaction, 36
traveling front solution, 36
Froude number, 61, 68
Fully hyperbolic problem, 181
Function
Bessel, 81
blending, 215
numerical flux, 205
periodic, 16
window, 30
FV approach, 183

G

Galerkin approach, 191
Gaussian, 120, 121
Geostrophic
approximation, 125
balance, 67, 99, 124
flow, 125
Geostrophy, 66, 125
Gibbs phenomenon, 223
Godunov method, 173

246 Index

Godunov-type finite volume method, 183, 188
Gravity
    acceleration due to, 54
    reduced, 77
    wave, 66, 68, 76, 124
Group speed, 94
Growth, 32
    law, 33, 35
    logistic, 14, 25, 35, 36, 47
    rate, 15, 26
Gyrotactic motion, 158, 162
Gyrotaxis, 157

H

Hardening spring, 13
Horizontal modes, 137
Hovmoeller plot, 33, 41
Hydraulic conductivity, 235
Hydrostatic, 57
    pressure, 59, 134
      relation, 55
Hyperbolic, 188
    systems, 188

I

Idealized internal seiche, 134, 138
Incompressible, 64, 227
Inertial oscillation, 76
Initial
    condition, 18, 23
      sensitivity to, 41
    perturbation, 38
    state, 37
    values, 23
Instability, 210
    barotropic, 89
    problem, 85
Integration time, 15–17
Interacting tracers, 36
Interior penalty (IP), 200
Internal seiche, 136, 142, 147, 221
Interpolation, 145
    transfinite, 216
Inverse scattering, 92
Irrotational, 64
    vortex, 64

J

Jacobian, 191, 192, 197, 198, 217
Jet
    double, 66

K

KdV equation, 92
Kelvin waves, 72, 76, 124
Kernel
    averaging, 120
    filtering, 120
Kinematic
    boundary condition, 58
    viscosity, 56
Kinetic energy (KE), 11, 12, 125, 126, 129

L

Lab frame, 52
Lagrange polynomial, 189
Lagrangian, 52
    measurement, 52
    transport, 139, 140
Laplace's equation, 211, 212, 227
Large eddy simulation (LES), 115, 120, 123, 126
Law
    conservation, 188, 202
    constitutive, 56
    Darcy's, 227, 229
    exponential, 32
      growth, 35
    growth, 33, 35
    Newton's second, 10, 12, 53
Lax-Friedrichs flux, 185, 190, 205
Legendre-Gauss-Lobotto (LGL), 194
LES, 126, 127, 129, 130
Limit cycle, 19, 20, 22, 23, 38, 155
    stable, 18, 20
Linear, 24
    advection, 207
      equation, 166
    models, 31
    momentum, 53
    Riemann problem, 203
    theory, 61, 83
    tilt, 65, 66
Linearize, 15
Local discontinuous Galerkin (LDG), 201
Logistic
    equation, 35
    growth, 14, 25, 35, 36, 47
Lotka-Volterra, 18
    model, 15–18, 23, 25, 28, 36, 37, 49
      parameters, 22
    system, 16, 18, 38

M

Mass matrix, 191, 195, 196, 217

Index **247**

Material derivative, 34, 52, 55, 116
Mathematical biology, 1, 4, 31, 32, 237
Matric potential, 235
Matrix
  flux Jacobian, 179
  mass, 191, 195, 196, 217
  stiffness, 191, 195
  Vandermonde, 193, 194, 197
Mechanical energy, 13
Mechanics
  classical, 28
Mesh generation, 220
Method
  collocation, 191
  discontinuous Galerkin, 184, 209
  finite volume, 166, 188, 202
  Fourier spectral, 33
  Godunov, 173
  numerical, 4, 91, 92, 165, 202, 205
  penalty, 200
  pseudospectral, 144
  SIP-DG, 201
  spectral, 101, 111, 126, 149
  time-stepping, 205
  unstructured grid, 190
  weighted residual, 191
  windowing, 30
MKS, 26
Modal, 192, 193
  filter, 199, 206, 209, 213
    filtering, 223
Model, 22
  advection diffusion population, 46
  advection-reaction-diffusion, 39
  basic, 3, 9
  birth-death, 156
  coupled, 26, 27
  dispersive shallow water, 92
  exponential, 32
  fluid mechanics, 26
  linear, 31
  Lotka-Volterra, 15–18, 23, 25, 28, 36, 37, 49
  nonlinear, 22
  parameters, 23
  plankton, 25, 26, 39
  predator-prey, 15
  shallow water, 199
  Smagorinsky, 123, 128
  wave, 91
  ZP, 39
Modeling
  mathematical, 22
Modes, 141

  horizontal, 137
  normal, 73
  vertical, 137
Modes of oscillation
  free, 65
Molecular diffusion, 117
Momentum equations, 59–61, 67, 71, 72, 93, 111, 179, 199, 206

N

Navier-Stokes, 118
  equations, 56, 116
Newtonian fluid, 56
Newton's second law, 10, 12, 53
Nodal, 189, 190, 192, 193
  approach, 195
Non-diagonal problem, 178
Non-hydrostatic pressure, 200
Nondispersive nature, 61
Nonlinear, 13, 15
  effects, 33, 34
  model, 22
  nonlinearity, 16
  quadratic term, 34
  wave train, 149
Nonlinearity, 16, 151
  quadratic, 36
Normal mode, 73
  form, 73
Normal vector, 55
Number
  Froude, 61, 68
  Rossby, 67
Numerical
  code, 32
  diffusion, 173
  dispersion, 126
  dissipation, 126, 128
  experiment, 18, 20, 21, 23, 31
    experimentalists, 19
    experimentation, 20, 21
  flux, 188, 190, 205
    function, 205
  integration, 26, 29
  methods, 4, 91, 92, 165, 202, 205
  simulation, 38, 65, 96, 159
  solution, 15, 37, 66, 212, 213, 235

O

Ockham's razor, 91, 118
Orbit, 16, 20, 24
Order of magnitude, 37

248 Index

Ordinary differential equation (ODE), 10, 13, 17, 20, 23, 33, 38, 54
  first order, 36
  model, 38
  solvers, 37
Oscillation
  inertial, 76
  relaxation, 16
  spatial, 33

P

Parameter
  biophysical, 33
  Coriolis, 221
  filtering, 199
  of convenience, 40
  regime, 23
  set, 40
  space, 26, 39
Parseval's theorem, 30
Partial differential equations (PDE), 32, 33, 69, 166
Passive tracer, 82, 83, 139
Pathline, 52, 62, 155
Penalty method, 200
Periodic, 16, 23, 30
  boundary conditions, 37
  functions, 16
  period, 23
  solution, 18
Permeability tensor, 227
Perturb, 38
Perturbation, 41
  initial, 38
  large, 41
Phase plot
  phase plane plot, 16
  phase space plot, 17, 28
  portrait, 19, 22, 24
Phase speed, 94
Phenomenon
  Gibbs, 223
  Runge, 223
Physical space, 33
Phytoplankton, 25, 26
  population, 25–27
Pinehurst Lake, 220
Plane wave, 32, 65
Plankton, 25
  models, 25, 26, 39
  phytoplankton, 25
    population, 25, 26
  zooplankton, 25, 26
    death rate of, 26

population of, 26
Planktor, 157
Poincaré waves, 70, 80, 100, 124
Point-wise product, 198
Poisson-type problem, 181
Polynomial
  characteristic, 70
  interpolation nodes, 194
  Lagrange, 189
Population, 4, 13, 32, 38
  evolution, 31
  growth, 13, 32
  predators, 15, 23, 38
  prey, 15, 20
  zooplankton, 25, 26
Pore fraction, 225
Pore space, 225–227
Position, 9, 51
Potential
  matric, 235
  velocity, 211
Potential energy (PE), 11
Potential flow, 64, 211, 232, 233
  solution, 212
  theory, 95, 212
Power spectral density, 21, 24, 29
Predation, 15, 22, 26
  by zooplankton, 25
  efficiency parameter, 22
  interaction, 38
  rate, 22, 26
  term, 18
Predator, 15, 16, 19, 22, 28, 38
  population, 15, 23, 38
Predator-prey model, 15
  predator-prey system, 15
Pressure, 55, 57, 229
  fluid, 54
  hydrostatic, 59, 134
  non-hydrostatic, 200
  water, 227, 234, 235
Prey, 15, 16, 19, 20, 22, 23, 38
  population, 15, 20
Problem
  adjustment, 95, 96
    rotating, 97
  elliptic, 181
  fully hyperbolic, 181
  instability, 85
  non-diagonal, 178
  Poisson-type, 181
  Riemann, 190, 200, 203, 205
    linear, 203

Index **249**

stability, 86
tilted free surface, 82
turbulence closure, 117
Pseudospectral method, 144

Q

Quadrature, 210, 217, 218, 223
Quasilinear form, 178

R

Random variables, 116
RANS, 57, 120, 123, 131
Rate of change
in time, 37
Rate of consumption, 15
Rate of strain, 123
Reaction, 31, 36, 37
advection equation, 32
diffusion, 31
equation, 32
terms, 37
Reduced gravity, 77
assumption, 77
Reductionism, 32
Relation
dispersion, 69, 70, 75
hydrostatic pressure, 55
Relaxation oscillation, 16
Restoring force, 12, 36
Reynolds
averaged, 57
averaging, 115, 120, 121, 130
decomposition, 119
stress, 119, 122, 130
Reynolds Averaged Navier Stokes (RANS), 120
Richards equation, 235
Riemann problem, 190, 200, 203, 205
Riemann solvers, 188
Rossby number, 67
Rotating adjustment problem, 97
Rotating seiche, 211
internal, 219, 220
Rotation, 66
Earth's, 62
local, 62
modified gravity waves, 70
Runge phenomenon, 223
Runge-Kutta
scheme, 19
solvers, 37

S

Saturated flow, 227

Saturation, 234
Scalar approach, 205, 206
Scheme
finite volume, 170
Runge-Kutta, 19
Seiche, 102, 230
barotropic, 104
breakdown, 107
development, 105
evolution, 102
internal, 136, 142, 147, 221
idealized, 134, 138
rotating, 211
internal, 219, 220
Series
Fourier, 28
time, 28–30
Shallow water
equations, 57, 60, 64, 66, 92, 115, 124, 125
linearized, 60
model, 199
speed, 61, 94
theory, 61
waves, 61
Shear
dispersion, 43
force, 55
simple, 62
stresses, 57
Shock, 92
Simple harmonic oscillator (SHO), 9–11, 18, 28, 31
Simulation, 82, 85, 127
large eddy, 115, 120, 123
numerical, 38, 65, 96, 159
Sink, 117
Sinking, 153
Slow-fast systems, 124
Smagorinsky
eddy viscosity, 124, 131
model, 123, 128
Solitary wave, 92, 206
Solution
DG-FEM, 210
numerical, 15, 37, 66, 212, 213, 235
periodic, 18
potential flow, 212
Solver
Riemann, 188
Runge-Kutta, 37
Source, 117
Space
Fourier, 33, 121
parameter, 26, 39

250 Index

physical, 33
pore, 225–227
Space-time, 33
Spatial
  derivatives, 37
  dimension, 33
  oscillation, 33
  variable, 37
Spectra, 20, 21, 24, 28–30
  computing, 28
  spectral, 21
    analysis, 24
Spectral method, 101, 111, 126, 149
Speed
  baroclinic, 77
  barotropic, 77
  convergence, 183
  group, 94
  phase, 94
  shallow water, 61, 94
Spline, 214, 219
Spurious eddies, 210, 213, 214, 219, 220
Spurious separation, 210
Stability, 15, 66
  problem, 86
Stable equilibrium, 14, 20, 22, 23
Stable limit cycle, 18, 20
Steady state, 15, 21, 26
  stable, 19, 35
  unstable, 35
Stiffness matrix, 191, 195
Stokes' theorem, 62
Stratification, 77, 92, 134, 143, 145, 147, 149
Stratified fluid dynamics, 133
Streamfunction, 45, 62, 64, 135, 141, 143
Streamline upwind Petrov Galerkin (SUPG), 184
Stress
  Reynolds, 119, 122, 130
  shear, 57
  tensor, 55
  viscous, 119
Subcritical flow, 61
Supercritical flow, 61
Surface integral, 196
Swim, 19
Swimming, 155
Symmetric interior penalty discontinuous Galerkin
      (SIP-DG) method, 201
System
  dispersive shallow water, 181
  hyperbolic, 188
  Lotka-Volterra, 16, 18, 38
  slow-fast, 124

T

Taylor dispersion, 43
Tensor
  permeability, 227
  stress, 55
Theory
  Brinkman penalization, 236
  linear, 61, 83
  potential flow, 95, 212
  shallow water, 61
Tilted free surface problem, 82
Time
  average, 116
  dependence, 23
  scale, 23, 26, 40
  series, 28–30
Time-stepping method, 205
Total energy, 11, 30
Tracer, 42
  active, 31
  biological, 82, 165
  evolution, 43
  interacting, 36
  passive, 31, 82, 83, 139
  transport, 66, 83, 88
Traction, 55
Transfinite interpolation, 216
Transport
  Eulerian, 139
  Lagrangian, 139, 140
  tracer, 66, 83, 88
Traveling wave solutions, 36
Turbulence, 45, 151, 153
  closure problem, 117
Turbulent
  closures, 115
  drag, 60
  motions, 56
Turbulent kinetic energy (TKE), 131

U

Uniform distribution, 38
Units
  system of, 26
Unsaturated flow, 233, 234
Unstructured grid method, 190
Upwind flux, 169
Upwind-biased, 188
  flux, 171
Upwinding, 169

V

Vandermonde matrix, 193, 194, 197

Index 251

Variable
  auxiliary, 198
  bathymetry, 180
  Eulerian, 139
  random, 116
  spatial, 37
  vertical, 161
Velocity, 52, 54, 55
  field, 42, 43, 139, 211
  potential, 211
Vertical
  modes, 137
  variable, 161
  wavenumber, 138
Viscosity, 213
  eddy, 56, 115, 117–119, 122–124, 128
  fluid, 227
  kinematic, 56
Viscous stress, 119
Volumetric water content, 235
Vortices, 126
Vorticity, 62, 64, 125–127, 129, 159, 160, 210

W

Warp factor, 195, 215
Water pressure, 227, 234, 235
Wave, 36
  ansatz, 69
  continuity equation, 199
  dispersion, 188, 189, 197
  dispersive, 70, 183
  front, 125–127
  gravity, 66, 68, 76, 124
    rotation modified, 70
  Kelvin, 72, 76, 124
  models, 91
  plane, 32, 65
  Poincaré, 70, 80, 100, 124
  shallow water, 61
  solitary, 92, 206
Wavelength, 33
Wavenumber, 33
  across-channel, 75
  vertical, 138
Wavepacket, 33, 38
  envelope, 33, 38
  modulated, 41
Wavetrain, 95, 107, 109, 147, 150, 206
Weighted residual methods, 191
Well-balanced schemes, 180
Wind, 169
Window
  -ing methods, 30
  function, 30

Z

Zooplankton, 25, 26
  death rate of, 26
  population of, 25, 26
Zooplankton-phytoplankton (ZP) model, 39

Printed in the United States
by Baker & Taylor Publisher Services